Assessing Landscape Resilience

Danny Tröger

Assessing Landscape Resilience

Biogeomorphic Interactions, Risk Perception, and Planning Evaluation of Forest Plantations in Patagonia, Chile

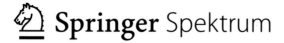

Danny Tröger
University of Kassel
Kassel, Germany

Doctoral thesis for the degree of
Doctorate of Natural Sciences (Dr. rer. nat.)
Faculty of Organic Agricultural Sciences
University of Kassel
Danny William Tröger,
born on 10.12.1985 in Marktredwitz
Witzenhausen, February 2024
Disputation 11. July 2024

ISBN 978-3-658-47273-3   ISBN 978-3-658-47274-0   (eBook)
https://doi.org/10.1007/978-3-658-47274-0

© The Editor(s) (if applicable) and The Author(s), under exclusive license to Springer Fachmedien Wiesbaden GmbH, part of Springer Nature 2024

This work is subject to copyright. All rights are solely and exclusively licensed by the Publisher, whether the whole or part of the material is concerned, specifically the rights of translation, reprinting, reuse of illustrations, recitation, broadcasting, reproduction on microfilms or in any other physical way, and transmission or information storage and retrieval, electronic adaptation, computer software, or by similar or dissimilar methodology now known or hereafter developed.
The use of general descriptive names, registered names, trademarks, service marks, etc. in this publication does not imply, even in the absence of a specific statement, that such names are exempt from the relevant protective laws and regulations and therefore free for general use.
The publisher, the authors and the editors are safe to assume that the advice and information in this book are believed to be true and accurate at the date of publication. Neither the publisher nor the authors or the editors give a warranty, expressed or implied, with respect to the material contained herein or for any errors or omissions that may have been made. The publisher remains neutral with regard to jurisdictional claims in published maps and institutional affiliations.

This Springer Spektrum imprint is published by the registered company Springer Fachmedien Wiesbaden GmbH, part of Springer Nature.
The registered company address is: Abraham-Lincoln-Str. 46, 65189 Wiesbaden, Germany

If disposing of this product, please recycle the paper.

Summary

This cumulative dissertation examines the resilience of landscapes with a particular focus on biogeomorphological interactions, risk perception and planning evaluation at the example of forest plantations in Chilean Patagonia. It examines the extent to which these plantations represent a suitable strategy to be resilient to disturbances such as climate change and natural disasters, while maintaining the essential functions, structures, processes and identity of the landscape over time, integrating three independent empirical studies with inter- and transdisciplinary approaches under the theoretical framework of landscape resilience.

The first study analyses how the interactions of vegetation and soil can make landscapes either more resilient or more vulnerable depending on land use and management. The concept of biogeomorphology is applied and a multivariate method is used to integrate pedological parameters and vegetation data. In the second study, the perception of natural hazards by the population is analysed. Qualitative empirical social research methods are used to construct a grounded theory. The results are discussed against the background of actor-network theory, including an epistemic reflection of the academic risk discourse. In a third study, the results of the first two research projects are incorporated into an evaluation of spatial planning. Here, the role of spatial planning in minimising risk and strengthening landscape resilience is examined and alternative approaches are identified. The work concludes with the identification of research gaps of supra-regional importance.

Zusammenfassung

Die kumulative Dissertation untersucht die Widerstandsfähigkeit von Landschaften mit besonderem Augenmerk auf biogeomorphologische Wechselwirkungen, Risikowahrnehmung und Planungsbewertung anhand des Beispiels von Forstplantagen im Chilenischen Patagonien. Es wird untersucht, inwiefern diese Plantagen eine geeignete Strategie darstellen, um gegenüber Störungen wie Klimawandel und Naturkatastrophen resilient zu sein und dabei die wesentlichen Funktionen, Strukturen, Prozesse und Identität der Landschaft über die Zeit zu bewahren. Die Arbeit integriert drei eigenständige empirische Studien mit inter- und transdiziplinären Ansätzen unter dem theoretischen Rahmen der Landschaftsresilienz.

In der ersten Studie wird analysiert, wie die Wechselwirkungen von Vegetation und Boden in Abhängigkeit der Landnutzung und dessen Management Landschaften entweder resilienter oder anfälliger machen können. Dabei wird auf das Konzept der Biogeomorphologie zurückgegriffen und eine multivariate Methode verwendet, um bodenkundliche Kennwerte und Vegetationsdaten integrativ zu betrachten. In der zweiten Studie wird die Wahrnehmung von Naturgefahren durch die Bevölkerung untersucht. Dabei kommen Methoden der qualitativen empirischen Sozialforschung zum Einsatz, um eine Grounded Theory zu konstruieren. Die Ergebnisse werden vor dem Hintergrund der Akteur-Netzwerk-Theorie diskutiert, einschließlich einer epistemischen Reflexion des akademischen Risikodiskurses. Dadurch soll ein dekolonialer Ansatz möglich werden. In einer dritten Studie fließen die Ergebnisse der beiden ersten Forschungsarbeiten in eine Evaluierung der räumlichen Planung ein. Hier wird die Rolle der Raumplanung bei der Risikominderung und Stärkung der Landschaftsresilienz beleuchtet und alternative Strategien vorgeschlagen. Die Arbeit schließt mit der Ableitung von überregional bedeutsamen Erkenntnissen und der Ableitung neuer Forschungsfragen ab.

Contents

1 Landscape Resilience at Risk—Forest Plantations as an Answer? 1
   1.1 Natural Hazard Regimes are Changing—Spatial Planning has to Adapt 1
   1.2 Theoretical Frameworks 2
      1.2.1 Landscape Resilience 2
      1.2.2 Risk 3
   1.3 Aims and Research Questions 4
   1.4 Rationales for the Selection of the Study Area 5
   1.5 Structure of this Book 7

2 Pinus Plantations Impact Hillslope Stability and Decrease Landscape Resilience by Changing Biogeomorphic Feedbacks in Chile 9
   2.1 Introduction 10
      2.1.1 Background 10
      2.1.2 State of the Art 11
      2.1.3 Open Questions 15
      2.1.4 Aims of the Study 16
   2.2 Study Site 16
      2.2.1 Physical Geography 17
      2.2.2 Soils 18
      2.2.3 Vegetation Assemblages and Land Use 19
   2.3 Materials and Methods 20

		2.3.1	Selection of Sampling Plots	20
		2.3.2	Derivation of Soil Indicators as Proxy Variables for Mechanical and Hydrological Mechanisms of Soil Stabilization	22
		2.3.3	In-Situ Measurements	23
		2.3.4	Laboratory Methods	24
		2.3.5	Statistical Treatment	24
	2.4	Results		25
		2.4.1	Soil Indicators Under the Different Vegetation Assemblages	27
		2.4.2	Multivariate Analysis of Soil Stability	29
	2.5	Discussion		32
		2.5.1	Vegetation Assemblage Specific Biogeomorphic Interactions and Hillslope Stability	32
		2.5.2	Impacts on Biogeomorphic Feedbacks	35
		2.5.3	Land Use, Biogeomorphic Feedbacks, Risks and Consequences for Landscape Resilience	39
	2.6	Conclusions		40
	2.7	Annex		41
3	*"Industry Impacts More Than Nature"*: **Risk Perception of Natural Hazards in More-than-human Worlds**			43
	3.1	Introduction		44
	3.2	State of the Art Theoretical Design		45
		3.2.1	Conceptualizations of Risk and its Perception	45
		3.2.2	Recent Critics and Potentials of the Actor-Network-Theory in Disaster Risk (Perception) Research	48
	3.3	Study Site		50
		3.3.1	Physical Geography	51
		3.3.2	Natural Hazards and Environmental Degradation	51
		3.3.3	Human Geography and Socio-Environmental Conflicts	52
		3.3.4	Sociological Lense: Vulnerabilities	53
		3.3.5	Disaster Event in 2007	54
	3.4	Materials and Methods		55
		3.4.1	Risk Map	55
		3.4.2	Risk Perception	55
	3.5	Results and Discussion		58

		3.5.1	Risk Analysis	58
		3.5.2	Risk Perception	62
		3.5.3	Hybrid Understanding for Symmetrical Relations	73
	3.6	Conclusions and Outlook		75

4 Navigating a Trojan Horse in the Last of the Wild: Pine Trees, Agroforestry, and Land Zoning Assemble the Landscape Resilience Dilemma in Patagonia 79
    4.1 Introduction 80
    4.2 Study Site 83
    4.3 Materials and Methods 86
    4.4 Results 88
        4.4.1 Goals, Objectives and Action Plans of the Regional Territorial Plan 88
        4.4.2 Disaster Prevention Plan 89
        4.4.3 Land Use Zoning 90
        4.4.4 Conformance Evaluation 92
        4.4.5 Performance Evaluation 93
    4.5 Discussion 94
        4.5.1 Low Performing Biodiversity Conservation and Natural Hazard Mitigation 94
        4.5.2 Neglected (Indigenous) Risk Perception 95
        4.5.3 Inspirations from Polycentric Governance 96
        4.5.4 From Critique to Proposals 98
    4.6 Summary and Conclusions 102

5 What Can We Learn from Aysén?—Discussion and Derivation of Follow-up Research Questions 105
    5.1 Biogeomorphic Feedbacks, Land Use and Landscape Resilience—Underestimated Interactions? 105
    5.2 Risk Perception: Does the Distinction Between Natural Hazards and Environmental Degradation Still Make Sense? Are We Looking at the Wrong Side of Risk? 106
    5.3 Reflection on the Contribution of Land Zoning to Landscape Resilience and Governance Analysis Research Proposal 107
    5.4 Lessons Learned from Chilean Patagonia 108

References 109

List of Figures

Fig. 2.1	Biogeomorphic feedbacks and landscape resilience	15
Fig. 2.2	Map of the study site	17
Fig. 2.3	Site properties, grouping by vegetation assemblage	21
Fig. 2.4	Exemplary soil profiles	25
Fig. 2.5	Boxplots of the soil stability indicators under the different vegetation assemblages	26
Fig. 2.6	Ordination plots (NMDS)	30
Fig. 2.7	Schematic overview of land use impacts on biogeomorphic feedbacks	36
Fig. 2.8	Boxplots of the soil properties (independent variables)	41
Fig. 2.9	Boxplots of the vegetation properties	42
Fig. 3.1	Topographic Map of the Study Site	50
Fig. 3.2	Risk Map	59
Fig. 3.3	Visualization of the grounded theory of risk perception	63
Fig. 4.1	Graphical abstract of the case study	82
Fig. 4.2	Map of the study area	84
Fig. 4.3	Map of preffered agroforestry zones and the locations of existing *Pinus* plantations	91

List of Tables

Table 3.1	Estimation of the number of persons at risk	60
Table 4.1	Overview of the operationalization of planning evaluation	87
Table 4.2	Categories of Risk Zones according to Greiving (2002, p. 275)	100

Landscape Resilience at Risk—Forest Plantations as an Answer?

1.1 Natural Hazard Regimes are Changing—Spatial Planning has to Adapt

Natural hazards regimes are changing worldwide due to climate and land use change (Gill and Malamud, 2017; IPCC, 2022; Kreibich et al., 2014; Ward et al., 2020). Current disaster events, like the flooding in the Ahr valley in 2023, demonstrate that these changes also affect Germany (Klose et al., 2016; Vorogushyn et al., 2022). It is evident, that the interactions between the natural events, such as heavy rainfall, and anthropogenic influences on ecological systems, like land use, become more and more relevant in the constitution of disaster risks (Gill and Malamud, 2017; Pedersen; Reinstädtler, 2023; Szymczak et al., 2022; Vorogushyn et al., 2022). Other land use related risks, like wildfires, are also likely to increase in Central Europe (Conedera et al., 2011; Feurdean et al., 2020). In many places, land use-related risks are exacerbated by a trend towards intensification of land use, while extensive, traditional land use is declining (Alaoui et al., 2018; Plieninger et al., 2020; Schils et al., 2022; van der Sluis et al., 2016).

Along with changing climate, areas of the world that had a low human footprint so far become more populated and deforestation increases there, such as in the cool-temperate and boreal forests of Canada, northern Europe and Siberia, but also in the southern hemisphere in Patagonia and New Zealand (Allan et al., 2017; van Well et al., 2018). With increasing settlement, natural hazards are becoming a more relevant issue there, for instance by the growing risk of wildfires (Holz and Veblen, 2011; Parisien et al., 2020). These forests are of enormous importance as carbon reservoirs for the global climate and their loss or inadequate reforestation threatens to further intensify climate change, which would probably lead to a further increase in meteorological hazards (Rockström et al., 2021; Windisch et al., 2023). An increasingly popular countermeasure is afforestation, often in the form of tree plantations with exotic species—though not without criticism (Ramcilovic-Suominen et al., 2022; Streck, 2020; Temperton et al., 2019).

Landscapes are the link between global climate change and specific local natural hazards and simultaneously are a suitable scale for analyzing risk relationships between different land uses (Johnson et al., 2023; Popp et al., 2014; Schaldach et al., 2020; Seppelt et al., 2013). Spatial planning can be a relevant component of disaster risk reduction at the landscape scale (Greiving, 2002; Kammerbauer, 2014). For several decades, engineering and technology-centered perspectives dominated this domain, which still has a formative effect today (Dombrowsky, 2008; Ueköter, 2018). Yet, the realization is gaining ground that nature based solutions and appropriate land use governance offer great potentials for disaster risk mitigation and spatial resilience (Kammerbauer, 2014).

Increasingly divergent and polarizing perceptions about the underlying causes of the risks by natural hazards create social tensions and intense political debates. The interweaving of spatial planning instruments for mitigation against natural hazards with land use conflicts calls into question if previous approaches to risk perception are helpful in spatial planning practice and land use governance (Christmann et al., 2014; Hurley and Walker, 2004; Kühne et al., 2021; Schneiderbauer et al., 2021; Truedinger et al., 2023).

This cumulative dissertation takes an empirical, inter- and partly transdisciplinary and exploratory approach to the question if afforestation with plantations is an adequate answer to the risks for landscape resilience, and the contribution of land zoning in this context. Using *Pinus* plantations in Chilean Patagonia as an example, at the center of the analysis are biogeomorphic interactions whose configuration can either mitigate, exacerbate or shift natural hazards, and the risk perception of local communities. The extent to which the results also provide impetus for the research field of landscape resilience, including as a problem field in spatial planning in Germany, is discussed in the final chapter and corresponding research questions for future works are derived.

1.2 Theoretical Frameworks

1.2.1 Landscape Resilience

As early as 2002, Greiving discussed disaster resilience as a fourth dimension alongside the economic, ecological and social dimension of sustainability in the context of spatial planning (Greiving, 2002, p. 203). As the term sustainability is

1.2 Theoretical Frameworks

meanwhile associated with an enormous number of definitions and contradictory discourses, the issues raised in Sect. 1.1 are not discussed under the heading of sustainable land use. Instead, the concept of landscape resilience is used, which—in the term itself—emphasizes the spatial scale of consideration and the purpose (Beller et al., 2015; Schmidt, 2022).

Landscape resilience as a concept and discourse builds on the works of Cumming (2011a) (eventhough called spatial resilience), and is extended by authors like Beller et al. (2015) and Schmidt (2022). Head (2012) further extended the perspective towards actor-network-theory and more-than-human-world approaches, overcoming the nature-culture-distinction in modernity (Latour, 2014, 1995, p. 36; Plieninger and Bieling, 2012). Based on the mentioned works, within this dissertation landscape resilience is understood as:

Landscape resilience is the capacity of a landscape to absorb disturbances or stressors, like climate change, natural disasters, or human activities, and maintain its essential functions, structures, processes, and identity over time. It involves building sufficient diversity and redundancy, resistance and elasticity, complexity and modularity, interconnectedness and self-sufficiency in its systems so that the landscape can recover and even evolve positively from perturbations. Landscape resilience is not just about maintaining the status quo but also about adapting to changing conditions and uncertainty, whilst minimizing negative impacts on ecological, social, and economic systems. It is a measure of the degree to which a landscape might be at risk of temporary or permanent loss of functionality following an impact.

1.2.2 Risk

The underlying understanding of risk and related terms such as hazard and vulnerability are based on the following discourses:

- the Pressure-and-Release model, defining risk as the intersection of hazard and vulnerability (Wisner et al., 2010),
- interrelations of vulnerability and environmental justice (Cutter, 2006),
- the riskscape concept to overcome constructivist and realist approaches (Müller-Mahn et al., 2018), and,
- the work of Neisser (Neisser, 2014), relating the riskscape concept and Actor-Network-Theory (Latour, 1995).

Further explanations to relevant theories and frameworks can be found in the corresponding subsections of section two, three and four.

1.3 Aims and Research Questions

The purpose of this work is to develop knowledge about risk relationships of forest plantations that spatial planning faces, and has to adapt to in the coming decades. The aim is to fill research gaps in the fields of landscape ecology and risk perception in this context. The guiding principle behind is to identify risk relationships and potential levers to achieve landscape resilience with spatial planning or closely related land use governance mechanisms (Chan et al., 2020; Fischer and Riechers, 2019; Martin et al., 2022). This results in four groups of aims and sub-goals, which are presented in the following four paragraphs and correspond to the Chaps. 2–5.

The first aim is to assess if the forest plantations in the study site can be capable to counteract erosion and landslides risks. Concepts are sought that reveal the interactions and feedbacks of land uses that influence the genesis of natural hazards or make the land uses more resilient to natural hazards affecting them. In particular, the focus will be on the interactions between soils and vegetation, i.e. inanimate and animate entities (Viles, 2020). These concepts could be helpful to integrate existing knowledge of ecological sub-disciplines and to analyze the consequences of different land uses for resilience at the landscape level, as well as to analyze synergies and trade-offs of natural hazard mitigation and nature conservation.

The second aim is to develop a broader understanding of risk perception, that can integrate non-western ontologies and overcome the separation of natural hazards and ecological degradation in the context of land use planning (Katanha and Simatele, 2019). Up to now, research on risk perception has been mainly oriented towards academic knowledge and the separation into disciplines, and strongly focused on the perception of hazards. As a result, planners were confronted with unexpected ontologies and discourses in practice (Gaillard, 2019b; Schneiderbauer et al., 2021). Thus, an approach on risk perception building on the Actor-Network-Theory will be developed.

The third objective targets a governance area that can contribute to landscape resilience: spatial planning. The question is raised as to what extent land zoning as one of its main instruments is adequately responding to the risks of erosion, landslides, wildfires and the if the risk perception of local communities is considered. Stumbling blocks are identified, and alternative strategies proposed.

In Chap. 5, there will be an overarching discussion of the findings and conclusions are drawn, to what extend the results are relevant to other regions, including Germany. New research questions that arised will be briefly outlined.

1.4 Rationales for the Selection of the Study Area

Why investigating risks for landscape resilience in Chilean Patagonia?

First, the ecological conditions are, with some limitations, comparable to those of Central Europe. The Aysén region is at a similar latitude to Central Europe and also has a temperate climate (Geiger and Pohl, 1961). However, in Chilean Patagonia large areas are still covered by primary forests, and the remaining areas are mainly extensively used (Hernández-Moreno et al., 2023). Under these conditions, the ecological interactions that are of interest in this work are likely to be more observable, as there are fewer disturbances. In contrast, in Germany land use has changed repeatedly over centuries, and therefore primary forests are scarce (Larsen et al., 2016).

Second, the region is subject to a spectrum of natural hazards, that are rarely present in Central Europe (Anacona et al., 2015; Gobierno Regional de Aysén, 2013; Oppikofer et al., 2012). It is likely that some of them will be present in Europe in the near future due to climate change (Feurdean et al., 2020; IPCC, 2022).

Third, the spatial planning system in Chile is currently being designed and implemented according to the German spatial planning system (Ossandón et al., 2020; Vicencio et al., 2023). Of course, there are major differences between the two state systems (and even more in the society), but some instruments, such as land zoning or the administrative division into sectoral competences and spatial planning as an cross-sectoral task, as well as the hierarchical division into municipalities, federal states and the national state, are comparable (Fürst and Scholles, 2008; Greiving, 2002; Vicencio et al., 2023). Conversely, findings from a planning evaluation in Chile can also be relevant for Germany, especially if they are sufficiently abstracted.

For these reasons, the Aysén region of Chile is a suitable study area for the research objectives described in Sect. 1.3.

In the risk perception study (Chap. 3), fishermen were interviewed, not residents or farmers near forest plantations. There are several reasons for this choice.

First it makes sense that the risks considered should be relevant to the repondents, e.g., that they have experienced them in the past. However, there are not an unlimited number of suitable, recent disasters available for empirical research. In 2007, the study area experienced a seismic crisis with hundreds of landslides. These occurred mainly on the coast and in the fjords west of the main Andean ridge and affected fishermen, but not farmers. These fishermen were therefore used as the study group for the risk perception study. Peasants' perceptions of the environmental risks posed by forest plantations and conflict potential have already been investigated (Braun, 2021b). Both study groups are included in the discussion on risk perception and the formation of the grounded theory (Chap. 3).

Second, we also wanted to represent indigenous perspectives. However, it is difficult to make contact with people in rural areas, including Mapuche or other indigenous communities, because their scattered farms are very difficult to reach and there are also guard dogs. A relatively large number of Mapuche fishermen live in Puerto Aysén, which made it possible to find suitable interviewees.

A third reason is that the local communities do not necessarily construct strict cause-and-effect relationships according to geoscientific knowledge of natural hazards. Consequently, it would be an inappropriate selection criterion to include only those participants in the study who are at risk from to a geoscientific perspective.

A fourth and closely related reason is that theories of risk perception have been developed in geographical areas where exposure to natural hazards is the exception. As a result, exposure has become a critical criterion for determining whether a risk exists. As a result, risk perception is usually only studied where people are actually geographically exposed. In Chile, however, natural hazards are almost omnipresent and Chileans, including the fishermen, generally assume that they are constantly exposed. The exposure criterion is therefore insufficient if the perception is actually taken seriously and a constructivist position is taken.

1.5 Structure of this Book

This book is essentially structured according to the questions formulated in Sect. 1.4. Each disciplinary topic corresponds to a published or submitted article of this cumulative dissertation, presented as Chaps. 2–4 in this book.

Chap. 2 will assess the ecological aspects and establishes the link to landscape resilience. Chap. 3 represents the study on risk perception. Both sections are incorporated in the planning evaluation presented in Chap. 4, which is the third article of the dissertation. There, the overarching question if forest plantations are an adequate answer to land use related risks in Patagonia is discussed and the role of land zoning critically assessed. As the articles were submitted to journals from various disciplines, the focus is in part strongly on current discourses in the respective disciplines. In Chap. 5, central results of all three article, or sections, respectively, are repeated once again if they are of supra-regional significance or are needed to draw conclusions about new research questions.

The initial hypothesis that forest plantations entails risks lies in the reduced plant biodiversity. It seemed plausible that the low plant diversity would also severely restrict other ecological functions relevant to landscape resilience (Braun et al., 2017b).

Pinus Plantations Impact Hillslope Stability and Decrease Landscape Resilience by Changing Biogeomorphic Feedbacks in Chile

2

This chapter is a reprint of the following article:
Tröger, Danny; Braun, Andreas Christian; Eichel, Jana; Schmidtlein, Sebastian; Sandoval Estrada, Marco; Valdés Durán, Ana (2022): Pinus plantations impact hillslope stability and decrease landscape resilience by changing biogeomorphic feedbacks in Chile. In: CATENA 216, S. 106364. https://doi.org/10.1016/j.catena.2022.106364.

Abstract

Forest plantations with exotic species are planted extensively in the southern hemisphere for soil conservation and shallow landslide mitigation. The extent to which these are suitable for fulfilling their protection goals is the subject of debate. A biogeomorphic framework was applied to link land use, soil conservation and natural hazards. It consists of a feedback loop with the two effect pathways *hillslope stability* and *vegetation fitness*. The study site is located in Chilean Patagonia, where thixotropic Andosols are widespread and *Pinus* plantations were planted initially in conservation areas, and later on private land.

We were testing the hypothesis whether soil stability differs between primary and secondary forests, *Pinus* plantations, wildfire sites (ex. plantations) and pastures. *Shear strength, liquid limit, consolidation degree* and *available water capacity* were used as soil stability indicators and set as dependent variables using non-metrical multidimensional scaling (NMDS), representing mechanical and hydrological biogeomorphic interactions. Soil texture, topographic and vegetation properties were *post-hoc* correlated as independent variables.

Vegetation assemblage correlates most strongly with soil indicator variance. The soils under secondary native *Nothofagus* forests have significantly higher *liquid limits* than *Pinus* plantations ($41 \pm 4.9\%$ vs. $31 \pm 14\%$, $p < 0.1$, A Horizon). *Consolidation degree* is higher under secondary forests than in Plantations (A and B Horizon), due to a significantly higher root abundance.

Primary forests provide landscape by maintaining the water cycle balance and biodiversity. Secondary forests establish the biogeomorphic feedback loop through mechanical effects and enhancing vegetation fitness. *Pinus* plantations cause a slight improvement in soil stability properties, but with trade-offs in water balance and vegetation fitness. Landscape resilience is thus impaired by the higher risk of wildfires, erosion and landslides. Pastures show good values in the soil stability parameters, but their biogeomorphic interactions are unlikely to rebuild landscape resilience.

2.1 Introduction

2.1.1 Background

Soil degradation and soil loss as a result of deforestation and subsequent land use is an ecological risk with local and global impacts (Foley et al., 2005; Ghazoul, 2013). Forests provide a multifunctional key-role especially in mountainous terrains, as they increase landscape resilience and hillslope stability by offering protection against erosion, landslides and maintaining the water cycle balance (Pawlik, 2013; Schmidt, 2022; Stokes et al., 2014). Therefore, efforts are being made in some regions to reforest, and this is often realized through forest plantations with exotic species (Bonnesoeur et al., 2019; DellaSala, 2020; Marden, 2012; Phillips and Marden, 2006; Zhang and Stanturf, 2008). This leads to a number of socio-ecological consequences, such as the impact on biodiversity, restrictions on use and sometimes even access by rural or indigenous communities, which in turn provokes conflicts with the owners of the plantations (Braun, 2021a; Braun et al., 2017b; Braun, (in Review); Brockerhoff et al., 2008; Gerber, 2011; Javier E. Gyenge, María Elena Fernández, Verónica Rusch, Mauro M. Sarasola, Tomás M. Schlichter, 2010; Malkamäki et al., 2018; Schirmer, 2007). We further argue that there are a number of unanswered questions regarding the success of this practice with respect to the nature conservation and natural hazard mitigation objectives themselves.

2.1.2 State of the Art

Many case studies address how afforestation and forest plantations contribute to slope stability. They highlight different aspects according to their perspective and methodology and therefore sometimes arrive at different assessments and recommendations for stakeholders, for example those involved in land use planning. These findings are roughly outlined in the following, before the current status and open questions of biogeomorphology follow, a framework which we expect will have an inheriting function.

2.1.2.1 Historical Development and Particular Findings

In Europe, erosion rates were highest after initial large-scale deforestation at the beginning of the Middle Ages, while later land uses, including forestry, re-stabilized slopes (Larsen et al., 2016). Tree species play an important role in this: Rickli et al. (2019) found that dense needle forests clearly prevent landslides, whilst gaps > 20 m essentially increase landslide susceptibility. In a case study in Japan it was shown that in the first 10 years after logging landslide probability is highest, whereas erosion rates are higher up until 45 years after logging (Imaizumi et al., 2008). Klaar et al. (2015) demonstrate the role of tree species in the evolution of deglaciated landscapes in Alaska in the shift of unstable to stable slopes of steep alpine terrain.

The mentioned studies show the crucial role of dense tree stands in shallow landslide and erosion risk mitigation by focusing on conifer trees (*Pinales/ Coniferales*) in their natural habitats. With the aim of erosion control and landslide risk mitigation, forest plantations with conifers have been and continue to be established in the southern Hemisphere, even though they represent non-native species for these ecosystems (Aburto et al., 2020; Bonnesoeur et al., 2019; Braun et al., 2017b; Farjon, 2010; Heaphy et al., 2014; Langdon et al., 2010; Page et al., 2000; Zhang and Stanturf, 2008). Here, the success of these measures seems to be less clear-cut, as explained below.

In a case study in New Zealand, it was shown that exotic pine plantations stabilize degraded hillslopes, which were susceptible to shallow landslides, gully erosion and earth flow movement (Marden, 2012). The authors used a GIS-based approach in which geomorphological inventory maps of erosion and landslide forms were generated. The causal processes are attributed to the interception of the canopy cover, the mechanical reinforcement by the roots and the reduction of the pore water pressure. Kamarinas et al. (2016) investigated the relationship between land use change (exotic forest plantation and pasture) and sediment load in the Hoteo River catchment in northern New Zealand at different temporal spatial scales, with exotic pine plantations being the most significant disturbance. The authors observed a clear correlation between increase in sediment load and deforestation, and conversely a decrease in sediment load after reforestation. In an intercontinental comparison between study sites in France and New Zealand, it was shown that while there is a consistent sequence of responses in channels and rivers to reforestation efforts and the direction of changes may be anticipated, the magnitude and timing of those responses are not (Phillips et al., 2013).

Exotic tree plantations are also excessively planted in Chile with the aim of erosion control and landslide risk mitigation (Braun et al., 2014; Braun, (in Review); Braun and Vogt, 2014; Carretier et al., 2018). In the coastal range of south-central Chile with a Mediterranean climate, Aburto et al. (2020) found that forest plantations suffer soil losses up to four times greater than native forests. In an climatically and ecologically comparable study area (south-central Chile), Banfield et al. (2018) found by applying ^{137}Cs radioisotopes as erosion proxies, that tree plantations might promote soil erosion—instead of preventing it. However, the harvesting method plays an important role in the effectiveness of the efforts on hillslope stabilization by reforestation with tree plantations (Carretier et al., 2018).

Forest plantations are also being established in Patagonia (southernmost part of America, on Chilean and Argentinean territory) with much cooler climates to counteract soil loss through erosion and landslides (Braun et al., 2017b; Langdon et al., 2010). Here, too, the success of these measures is the subject of controversy. In Argentinian Patagonia, for example, the uppermost 15 cm of soil has been lost within the last 50 years in 70 % of sample points under pastures, against 58 % in plantation and 45 % in native forest sampling points, as La Manna et al. (2019) determined on basis of ^{137}Cs inventories. However, soil degradation expressed via physical, chemical and magnetic soil properties differs only slightly between pasture and forest plantations and a higher proportion of organic matter is found under pastureland (La Manna et al., 2018). Nevertheless, *Pinus* plantations show higher erosion control than pasture in rainfall experiments (La Manna et al., 2016).

2.1.2.2 Biogeomorphic Interactions, Feedbacks and Landscape Resilience

The mentioned studies commonly consider the contribution of afforestation to either erosion *or* landslides. However, ecological mitigation of hillslope instability should consider both risks (Stokes et al., 2014). Apart from this, the above mentioned case-studies assess plant-soil interactions in a unidirectional way, which, however, does not adequately reflect reality (Amundson et al., 2015; Stallins, 2006). To address these criticisms, biogeomorphic frameworks, concepts and principles were established and are further developed (Corenblit et al., 2011; Marston, 2010). To describe plant-soil interactions which stabilize soils and thus the earth's surface, biogeomorphology uses ideas, components and principles from system theory such as reciprocity, path dependence, nonlinearity, complexity, and so on (Eichel et al., 2013). Applications of these frameworks have been conducted and continuously extended to understand the role of forests on hillslopes for soil creep, biomechanical aspects and landscape dynamics (Pawlik et al., 2016; Pawlik et al., 2013; Pawlik and Šamonil, 2018). Biogeomorphic feedback systems have also been studied on para- and periglacial environments, demonstrating the role of vegetation on lateral moraine slope stabilization, as well as their role in wildfire regimes (Eichel et al., 2018; Eichel et al., 2016; Stallins, 2006; Stine, 2016).

Biogeomorphic feedbacks and their importance for hillslope stability and landscape resilience are schematically shown in Fig. 2.1. One side of the feedback loop is the improvement of hillslope stability by vegetation, which consists of landslide mitigation, erosion protection and water cycle balance maintenance, and the other side is the increase of vegetation fitness via improving conditions for plant growth, reproduction and survival (Eichel et al., 2017). Both sides together increase landscape resilience, which is the "adaptability and self-renewal capability of a landscape and thus the ability of a landscape to maintain, renew and strengthen its own fundamental landscape qualities despite disturbances, crises or ongoing changes" (Schmidt, 2022, p. 6). Landscape resilience maintains a number of protective functions, like regulating the water cycle and natural wildfire regimes, enabling pedogenesis and maintaining the nutrient cycle, and, last but not least, the mitigation of erosion and landslides. Of course, landscape resilience involves much more, but ultimately behind it all is the question: how can the land use on hillslopes be optimized not only for benefits on the site itself, but how do regional synergies arise for other uses and stakeholders (Buma et al., 2019; Dawley et al., 2010; Hudson, 2010; Schmidt, 2022; Walker et al., 2012).

The biogeomorphic feedback loop itself is the result of a whole series of ecological processes, many of which are plant-soil (or: biogeomorphic) interactions. On the one hand, these are hydrological processes, such as interception, transpiration, infiltration, etc., and on the other hand processes in the soil. In order to classify the vegetation effects relevant for hillslope stability, they can divided into mechanical and hydrological mechanisms, respectively (Greenway, 1987, modified by Sidle and Ochiai, 2006, cited in: Marston, 2010). Slope-stabilizing mechanical vegetation effects include, in particular, anchoring by roots, through the established connection with deeper soil horizons and between weak layers and support of upslope soils (Balzano et al., 2019; Marston, 2010). In addition, trees contribute directly to geomorphologic processes, for example, by trapping downslope sediments on the trunk (accumulation of particles from the clay fraction to boulders) or in pits of uprooted trees (tree uprooting) (Pawlik, 2013). Other mechanical effects of vegetation are changes in soil properties, for example, the increase of shear strength through biomechanical and biochemical weathering (Marston, 2010; Pawlik, 2013). The hydrological mechanisms are vegetation-specific, partly due to above-ground structural differences, for example, the throughfall kinetic energy of raindrops, and partly indirectly due to processes in the soil, for example specific transpiration rates and hence soil moisture levels (Marino et al., 2020; Marston, 2010; Senn et al., 2020; Wicki et al., 2020). Especially from soils derived from volcanic ash (Andosols), it is known that structurally relevant soil parameters, soil pore processes and with it the water storage are influenced by land use related mechanical stress and wetting and drying cycles (Dec et al., 2012; Dörner et al., 2009).

2.1 Introduction

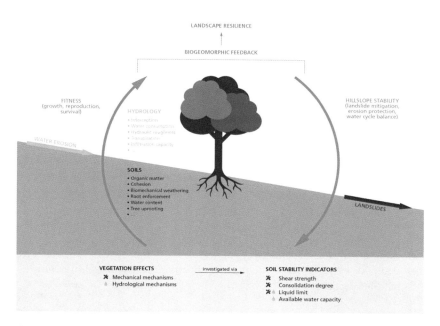

Fig. 2.1 Biogeomorphic feedbacks and landscape resilience. (Concept: Eichel. Graphical design: Gesa Siebert (Institute of Regional Science, KIT, Karlsruhe))

2.1.3 Open Questions

Biogeomorphology provides a suitable framework for investigating the extent to which forest plantations are suitable for fulfilling their original objectives of erosion control and landslide risk mitigation. Nevertheless, there are a number of unanswered questions, if only because different trees behave in different ways and the impact of native and invasive species might differ (Viles, 2020). In a systematic review Viles (2020) concludes that "a key question (…) today is whether manipulating biogeomorphic interactions can lead to more resilient landscapes and social-ecological systems and if so, under what conditions. (…) A key research need for the future is to explore the multiscalar characteristics of adaptive cycling within biogeomorphic systems and investigate how biogeomorphology (or: geomorphology and ecology, authors note) might contribute further to environmental management, conservation and risk reduction." The corresponding findings could provide for "practical application of understanding of multi-scalar, adaptive cycling in biogeomorphic systems to environmental management, conservation and risk reduction" (Viles, 2020).

2.1.4 Aims of the Study

According to the line of argumentation just quoted, this study aims to investigate the impact of land use on soil properties relevant to hillslope stability and erosion, with special emphasis on *Pinus* plantations with exotic species and secondary forests with native species. In addition to classical parameters of soil stability, those relevant to biogeomorphic interactions and feedbacks are also investigated. The research objectives are therefore:

- Assessment of soil indicators under different vegetation assemblages that:
    - express soil stability directly
    - influence hillslope stability via soil properties and conditions for biogeomorphic feedbacks

- Multivariate analysis of soil stability indicators with both site and vegetation characteristics as independent variables
- Evaluation of the contribution of exotic forest plantation regarding soil loss conservation and ecological landslide risk mitigation in comparison to secondary native forests
- Establish the connection of vegetation assemblage, land use, nature conservation and natural hazard mitigation using a biogeomorphic framework

2.2 Study Site

The study site (Fig. 2.2, 45°34′ S, 72°04′ W, 300 m a.s.l.) is located in the Chilean part of Patagonia around Coyhaique, the Capital of the Region Aysén.

2.2.1 Physical Geography

Coyhaique lies at about 300 m above sea level to the east of the main Andean ridge, while the two mountains bordering the valley to the north and south reach altitudes of up to about 1500 m above sea level. Formed by glaciers during the Pleistocene, the relief is today characterized by U-valleys and moraines with hanging V-valleys cut into them (Glasser et al., 2008). A boreal climate prevails in the study area, which is assigned to the type CfB according to the effective climate classification by Köppen/Geiger (Geiger and Pohl, 1961). The average annual temperature is approx. 4.8 °C and the annual precipitation is 1589 mm (climate-data.org, 2021).

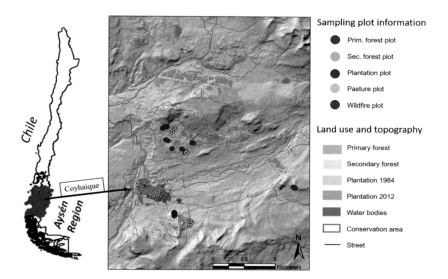

Fig. 2.2 Map of the study site. (Land use and sampling plots as well as presentation of the increase of Pinus plantations between 1984 and 2012. Cartography: Tröger. Supervised Classification: Tröger. Remote Sensing Data: ASTER GDEM; LANDSAT 5 TM (NASA, 1984) and LANDSAT 7 ETM + (NASA, 2012). GIS Data: Albers, 2012)

2.2.2 Soils

The soil substrate in the study area mainly consists of volcanic ashes, as well as glacial and glaciofluvial sediments (Casanova et al., 2013; Sandoval E., 2014; Stolpe et al., 2014). The soils are assigned to the Andosols (Stolpe et al., 2014, 57 f.). They show a striking stratification where several tephra falls deposited over the initial parent material (Casanova et al., 2013, 70 ff.). The soils found in the study area have a low bulk density between 0.6 and 1.1 g cm^{-3} (Sandoval E., 2014, p. 108). The proportion of the clay fraction is generally very low and typically increases in the deeper horizons (Casanova et al., 2013, p. 71; Sandoval E., 2014). Soil particles of the clay fraction in Andosols consist to a significant extent of allophanes, which is why the soils are usually thixotropic (Basile et al., 2003; Dexter, 1988). This special property is related to variations in the arrangement of the particles and changes in the absorbed water (Dörner et al., 2012). The organo-mineral complexes can vary from a rigid and resistant state to a more fluid consistency (Porta et al., 2013). The cohesion of andic soils can change with environmental conditions, for example, when the water content changes or mechanical disturbances have altered the structure once (Maeda et al., 1977). Accordingly, soil creep processes and landslides are typical for volcanic ash soils, with the failure plane often being the boundary to buried horizons (Basile et al., 2003; Maeda et al., 1977; Sandoval E., 2014). Around Coyhaique, the ash layers lie over alluvial terraces and form the Pollux series (Casanova et al., 2013, p. 74). Due to the widespread occurrence in Patagonia, soils of the Typic Hapludands were selected for the study. In terms of pedogenesis, the soils were homogeneous until the beginning of human intervention about 120–140 years ago (Azócar García et al., 2010; Bizama et al., 2011; Ibáñez Santa María, 1973). A general description of the horizon sequence and chemical soil parameters of the soils under different land uses is presented in [Change compared to the published article; author's note] the appendix and supplementary material of the digital version [End of the Change; authors note].

2.2.3 Vegetation Assemblages and Land Use

The native forests in the study area are formed by trees of the genus *Nothofagus* (Veblen, 1996). The climatic gradient (drier eastwards), tree canopy cover and grazing intensity are the main factors of the species composition within the forests (Alonso et al., 2020). Volcanic eruptions, landslides and windthrows are the main natural disturbances and consequently responsible for forest regeneration, whilst wildfires hardly played a role in the native succession (Fajardo and Gundale, 2015; Haberzettl et al., 2006). However, wildfires gained significance as vegetation disturbance factor since the beginning of anthropogenic activity in Patagonia at the end of 19th Century and especially with the intensification of land use since the 1920th (Ibáñez Santa María, 1973; Markgraf et al., 2007; Paritsis et al., 2015). Before colonization by settlers from more northerly parts of Chile and from Europe, there were indigenous groups living nomadically in Patagonia, but on a macro scale they had little influence on the vegetation (Aguilea and Tonko, 2013; Ibáñez Santa María, 1973).

In the study area, as well as in many parts of Patagonia, extensive and intensive land use was prepared by slash-and-burn clearing, mainly in the 1930 s and 1950 s (Fajardo and Gundale, 2015; Langdon et al., 2010; Sánchez-Jardón et al., 2010). These fires often got out of control and caused around 3 million hectars of forested land to be lost (Azócar García et al., 2010; Bizama et al., 2011). Overgrazing (sheep and cattle breeding) increased the subsequent erosion (Bonilla et al., 2014; Fajardo and Gundale, 2015). Meanwhile, livestock is unprofitable and takes place mainly in the form of subsistence farming, but grazing intensity still exceeds the ecological capacity around Coyhaique (Castellaro et al., 2016). As a result of last centuries land-use, soil erosion processes became one of the most obvious environmental problems and source of natural hazards in the region (Casanova et al., 2013, 71 ff.). Besides high erosion rates, soil loss occurs through shallow translational landslides.

In order to counteract soil erosion, reforestation with exotic species was begun in the 1960 s and 70 s, with test plantations in the Nature reserve "Reserva Nacional Coyhaique" (Braun et al., 2017b; Fajardo and Gundale, 2015; Langdon et al., 2010). Mainly *Pinus ponderosa* and *Pinus contorta* were planted, but also *Pinus sylvestris* and *Pseudotsuga menziesii* (Braun et al., 2017b; Fajardo and Gundale, 2015; Langdon et al., 2010). In recent decades, the area covered by *Pinus* plantations has increased significantly, especially on the hillslopes around Coyhaique (Fig. 2.2). Natural forest regeneration (Secondary forests) consists of very dense *Nothofagus* stands, in which multi-stem *Nothofagus pumilio* individuals with relatively low genetic variance spread from the edges of primary forests (Till-Bottraud et al., 2012).

To sum up, the anthropogenic ecosystem transformation of the last century led to a landscape mosaic, which can be schematized as following (Candan and Broquen, 2009; Fajardo and Gundale, 2015): Primary forests can be found on the higher slopes, towards the climatic tree line. In lower altitudes follow secondary *Nothofagus* forests resulting from natural regeneration or planted *Pine* plantations. In the valleys, Pastures with different intensity of use but also *Pine* plantations are typical, as well as field cultivation. Towards the east, where it becomes drier, forests and shrublands are followed by steppe ecotones.

2.3 Materials and Methods

2.3.1 Selection of Sampling Plots

A total of 29 plots were set around Coyhaique in soils of the Typic Hapludands (Fig. 2.2, Fig. 2.3, Sect. 2.2, [Change compared to the published article: Deletion of the reference to the annex. The data is attached digitally; authors note]). The selected plots are all located in areas for which the characteristic properties of this soil type can be assumed to be having been homogeneous before anthropogenic land use change, which started around 120–140 years ago. Differences in soil properties therefore are assumed to result mainly from land use changes. Within areas defined as representative in terms of pedogenesis and vegetation assemblage (see Sect. 2.2), exact locations for soil pits were then selected using the random walk method (Fajardo and Gundale, 2015; Underwood, 1997). Originally, forest plantation plots were established in sites with different *Pinus* species, but were later treated as one group (*Pinus* plantation) due to homogeneous properties amongst them.

2.3 Materials and Methods

Site properties
Grouping by vegetation assemblage

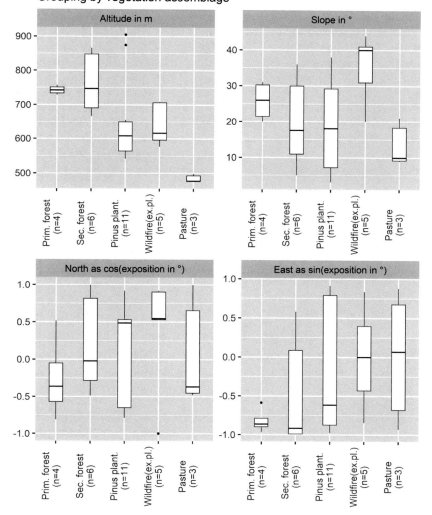

Fig. 2.3 Site properties, grouping by vegetation assemblage

2.3.2 Derivation of Soil Indicators as Proxy Variables for Mechanical and Hydrological Mechanisms of Soil Stabilization

The use of soil indicators is common for sustainability management (Boyle and Powers, 2013), whilst it seems to be less common when assessing erosion and landslide risk, as shown in the introduction. Most case studies whether measure sediment yield or observe geomorphological surface forms such as landslide scars. However, research designs should incorporate fundamental processes and their interactions (Sidle and Bogaard, 2016). For this purpose, soil indicators are suitable, as they express the susceptibility of soils for superficial soil erosion and shallow landslides (Costantini et al., 2016; Stokes et al., 2014). In this line of argumentation, the soil properties underlying the susceptibility to water erosion resistance and landslide risk will be expressed by the following four indicators:

- *Shear strength*, which expresses the maximum shear stress that the soil can withstand. Shear strength is used as a soil mechanical parameter in engineering geology for slope stability analysis; hence, in reality this property is often heterogeneous and difficult to determine in case of root penetration and the like (Genske, 2006). Shear strength is associated with mechanical mechanisms (see Sect. 1.2).
- *Consolidation degree*, expressing the cohesion of soil packages against mechanical stress at the macro level. It is representing the role of root reinforcement and consistency of root-soil structures (Genet et al., 2008; Hudek et al., 2017; Wang et al., 2020; Watson et al., 1999). Consolidation degree is associated with mechanical mechanisms (see Sect. 1.2).
- *Liquid Limit*, representing the ability of the soil to absorb water before changing the consistency from solid to liquid. This is crucial for the process of liquefaction on the mikro- (erosion) and makroscale (landslides) (Somos-Valenzuela et al., 2020b; Somos-Valenzuela et al., 2020a; Stanchi et al., 2015; Stanchi et al., 2013). *Liquid limit* is associated with mechanical and hydrological mechanisms (see also Sect. 1.2) (Basile et al., 2003; Dexter, 1988).
- *Available water capacity*, expressing the water storage capacity of the soil. It is calculated as the difference between the amount of water plants cannot move against gravity and the amount of water held in the soil against gravity (and therefore not running off downwards into deeper soil horizons or downslope). The water content of the soil, in turn, has a direct impact on various processes relevant for hillslope stability, for example due to the content of water

present in the soil during dry periods or in the event of a heavy rainfall event (Costantini et al., 2016; Marston, 2010; Sidle and Bogaard, 2016; Terwilliger, 1990). Available water capacity is associated with hydrological mechanisms (see Sect. 1.2).

The selected soil indicators represent proxy variables that play a role in both mechanical and hydrological mechanisms of soil stabilization, as outlined in the introduction (see also Fig. 2.1). Technical details can be found in the next two sections.

2.3.3 In-Situ Measurements

The field campaign with in-situ measurements and soil sample collection took place between May and June 2016. Vegetation was registered according to Braun-Blanquet (1928), estimating the *tree, shrub* and *herb layer coverage*. The *age* of the forest stands is determined based on tree rings, whilst the oldest tree of at least four replicates in each plot was assumed as the stock age. This is assumed to be the time period of soil property changes due to land use. The age of primary forests cannot be determined exactly, the values used for statistics represents the minimum age.

Soil parameters were taken in soil pits of 1 m depth according to the german pedological mapping guide (Eckelmann et al., 2006) in the A and B Horizon, before taking soil blocks for laboratory analyses. *Root penetration intensity* was differentiated in coarse and fine roots as the number of roots per 10 cm^2 in five classes (Eckelmann et al., 2006, p. 129).

The *shear strength* in kg·cm^{-2} was measured with the pocket shear stress measuring device *Pocket Vane Tester* with the CL 101 adapter. According to the operating instructions, the measuring range of this adapter with a diameter of 48 mm is 0 to 0.2 kg·cm^{-2} with a reading accuracy of 0.01 kg·cm^{-2}. At least 5 measurements per soil pit and horizon were taken in each case in freshly exposed areas that were free of roots as far as possible, and the arithmetic mean was calculated from these measurements.

The macro stability is measured via the consolidation degree expressed in 5 levels according to german pedological mapping guide (Eckelmann et al., 2006, p. 122): A block of soil is removed and dropped from a height of 1 m onto a hard surface (e.g. a spade) and the size, number and strength of the fragments is determined. Level 1 means that the soil already disintegrates into small aggregates during removal. Level 5 means the soil block retains its structure even after the impact on the ground and can hardly be reduced in size by hand.

2.3.4 Laboratory Methods

Soil samples of each A and B Horizons were analyzed in June and July 2016 at the soil laboratory of the University of Concepción, Facultad de Agronomía, Campus Chillan.

The *soil texture* in percentage of the total fine soil (determination of the soil type) was determined by sieving and the sedimentation method (Hartge and Horn, 2009, 28 ff; Sandoval E. et al., 2012, 21 ff.).

The *bulk density* in g•cm^{-2} was determined using the plunger cylinder method (Hartge and Horn, 2009, 43 ff; Sandoval E. et al., 2012, 38 ff.). The cylinders were taken from soil samples transported in boxes.

The *liquid limit* (consistency limit) according to Atterberg in weight percent (Scheffer et al., 2010, p. 195) is determined according to Casagrande (DIN 18122 Part 1).

The *available water capacity* in weight percent is calculated from the difference between the water capacities in weight percent at 300 hPa and 15000 hPA (negative and positive pressure method) (Hartge and Horn, 2009, 81 ff; Sandoval E. et al., 2012, 65 ff., 73).

2.3.5 Statistical Treatment

All statistical analyses are performed with the R software (RStudio Team, 2019). For multivariate analysis a non-metrical multidimensional scaling (NMDS) was performed, which ordinates the sampling plots due to (di)similarity in its dependent variables, in this case the soil stability indicators. This frequently used method in plant ecology can additionally relate environmental conditions, vegetation and soil indicators by a post-hoc correlation of independent variables. Either the vegetations' response to environmental and soil conditions can be analyzed, or soil properties can be considered as dependent on vegetation properties (Eichel et al., 2013; Kerns et al., 2003). In the present study, the sites were ordinated according to their (di)similarity of soil indicators as proxy variables for soil stability (see Sect. 3.2), using the Euclidean distance. For this purpose, the *metaMDS* method in R was applied after scaling and root transformation of the dependent variables. The post-hoc correlation of independent variables, in this case environmental variables and vegetation indicators (see Sect. 3.2, 3.3 and Fig. 2.9), was performed with the *envfit*-method (Oksanen et al., 2019; RStudio Team, 2019). For pairwise comparisons of particular indicators under different land uses the Welch-Test was used.

2.4 Results

Figure 2.4 shows one exemplary, characteristic soil profile of each considered vegetation assemblage. A general comparison of soil stability indicators in the A and B Horizon is shown in Fig. 2.5.

Fig. 2.4 Exemplary soil profiles. (Soils correspond to the soil profiles described Sect. 2.2 (Andosols, Typic Hapludands)

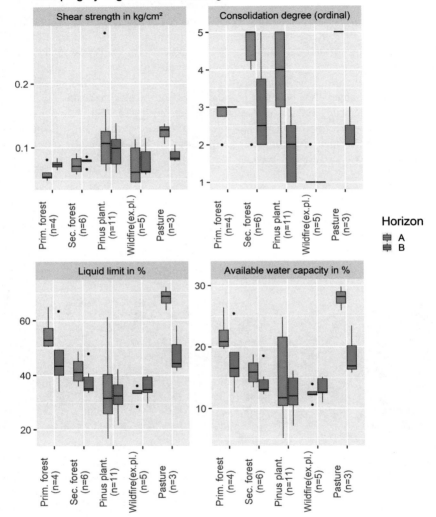

Fig. 2.5 Boxplots of the soil stability indicators under the different vegetation assemblages. (A Horizons in red, B Horizons in blue)

2.4.1 Soil Indicators Under the Different Vegetation Assemblages

In this section, specifics are highlighted that are particularly relevant to the research question and mandatory for interpretation, including impressions from fieldwork which are not represented in the pure numerical data. In fact, different aspects are highlighted for each vegetation assemblage. This seems confusing, but is necessary in order to then obtain an overall picture later on when the relevance for biogeomorphic feedbacks is established.

Primary *Nothofagus* forests: During fieldwork, the soils under primary *Nothofagus* forests were generally very moist. They exhibit a very low *shear strength* with little variance in both horizons, both in terms of absolute values and compared with the other vegetation assemblages. Their *consolidation degree* is in the middle range of absolute values with a median at level 3 with small variance for both horizons. The *liquid limit* is significantly higher than under *Pinus* plantations ($p < 0,1$) in both horizons. Regarding the *available water capacity,* primary *Nothofagus* sites have the highest values in both the A and B Horizon of all considered vegetation assemblages, with exception of the A Horizon of pasture soils.

Secondary *Nothofagus pumilio* forests: During fieldwork, some of the soils were difficult to penetrate for sampling due to their dense root systems. The *shear strength* of secondary *Nothofagus pumilio* forests in the A Horizon is significantly higher than under primary forests ($p < 0.01$), but lower than under *Pinus* plantations (even though not significant, $p > 0.01$). In the B Horizon, the same pattern can be observed. The *consolidation degree* of secondary forests in the A Horizons has its median in the most stable category 5, with a Minimum at 3. This means that more than 50 % of the soils retain their structure almost entirely even after massive mechanical stress. In the B Horizons, the median is a at level 2,5, meaning that 50 % of all plots had soils retaining their structure after mechanical stress at least partly. In both horizons the median of the consolidation degree of secondary *Nothofagus* forests is higher than under *Pinus* plantations. The *liquid limit* of the A Horizon is significantly higher than in *Pine* Plantation´s soils ($p < 0.1$). In the B Horizon the median is higher than under Plantations, yet not significant. The median of *available water capacity* under secondary forests is higher than under plantations in both horizons, yet not significant ($p > 0.01$).

Pinus plantations: The soils in *Pine* plantations, especially those under *Pinus contorta*, appeared very dry during fieldwork and some had an obvious, strong Mycorrhiza. *Shear strength* in A Horizons is significantly higher than under primary forests ($p < 0.1$) and also higher than under secondary forests, yet not significant ($p > 0.01$). In the B Horizon *Pine* plantations have the highest values of all considered land uses, but differences are not significant. In both horizons, the values are nevertheless in the lower range of the absolute scale.The median of *consolidation degree* in the A Horizon is at level 4, which is one level lower than among secondary forests. In the B Horizon, median of consolidation degree is at level 2 and therefore also below the levels of secondary *Nothofagus* forests. However, the variance is interesting: some soils under secondary forests have the most stable level 5 in the B horizon, while the levels 4 and 5 are not reached under Pinus plantations.The *liquid limits* in the A Horizons are significantly lower than among secondary native forests ($p < 0.1$). In the B Horizon median is also lower, but not significant. Remarkable is the fact, that the median of liquid limit of plantations is at comparable low levels like at the Wildfire plots. The *available water capacity* has the lowest median of all vegetation assemblages, but with a large variance especially in the A Horizon, where scattered high values also occur.

Wildfire sites (ex. Plantation): The *shear strength* of Wildfire sites lies at a comparable (low) level as primary forests, but with greater variance. The *consolidation degree* lies at level 1 in both horizons, representing the lowest possible values, without any outliers upwards. *Liquid Limits* lie at comparable levels like in *Pinus* plantations, but with less variance. The same applies to the *available water capacity*.

Pastures: The soils of the pasture plots had light soil frost at the time of sampling. The pasture sites have the highest *shear strength* values in the A Horizon A of all the assessed vegetation assemblages. The *consolidation degree* of the A Horizon is at level 5 (the highest level) without outliers downwards and therefore by far the most stable of all considered land uses. In contrast, in the B Horizon the Median is at level 2, representing comparable low levels as under plantations and Wildfire sites. The *Liquid limit* in the A Horizon is significantly higher than under all other vegetation assemblages. In the B Horizon values are comparable with those of primary forests. The same applies to the *available water capacity*.

2.4.2 Multivariate Analysis of Soil Stability

Figure 2.6 shows the similarities of the sites with respect to soil stability, plotted in the two-dimensional space of the first two principal axes of the NMDS. Sites with similar soil stability properties are close to each other. The sites are forming clusters according to the vegetation assemblage, which is made even more obvious by the enveloping ellipses.

Secondary forests lie closest to the primary forests both in the A and B Horizon. Therefore, their soil stability properties are more similar to the original ones. In the A Horizon this mainly concerns the properties *liquid limit* and *available water capacity* and in the B Horizon additionally the *consolidations degrees*. The *Pinus* plantations are located further away from primary forests than secondary forests. This is even clearer in the B Horizon, where there is no overlap with secondary forests. In the B Horizon the *Pinus* plantations are very close to the Wildfire sites, meaning their soil stability properties are comparable (weak).

The post-hoc correlation within the NMDS plot (Fig. 2.6) demonstrates which of the environmental and vegetation variables correlates most strongly with which indicator of soil stability (Significant correlations are drawn in red, $p = 0.001$).

Only few vegetation and environmental variables correlate significantly with the soil stability indicators. For the A Horizon these are: *Herb cover*, *Clay content*, *slope angle* and *vegetation assemblage* (variable name: lu*(*vegetation assemblage*)). Among them, vegetation assemblage shows the strongest correlation ($r^2 = 0.47$), demonstrating the crucial role of tree species composition for the development of stable soils. *Herb cover* correlates strongly with Pasture, which is not surprising as tree-covered land uses wouldn´t reach comparable levels of herb coverage in the understory than grasslands. *Clay content* on Plantations and Pastures is higher than average, whilst on Wildfire sites, *clay content* is lower (Fig. 2.8). The opposite is true for the variable *slope*, which correlates strongly with Wildfire sites. Both observations are due to the same cause: all sample plots of Wildfire sites are located at the same steep hillslope, as only here Wildfire occurred. The other vegetation assemblages have also sample plots in less steep terrain. The downward shift of the clay due to relief explains the pattern of shear strength between the vegetation assemblages (see Fig. 2.5). Hence, it is not strong enough to speak of a different soil type and, moreover, the values are at the lower end of the absolute scale and the differences have little effect on stability.

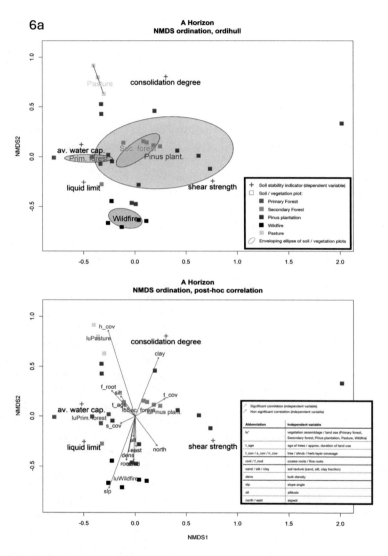

Fig. 2.6 Ordination plots (NMDS). Soil stability variables (dependent variables) with enveloping ellipse (standard deviation) and post-hoc correlation of independent variables. (6a: A-Horizon. 6b: B-Horizon)

2.4 Results

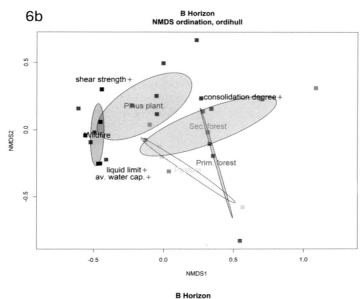

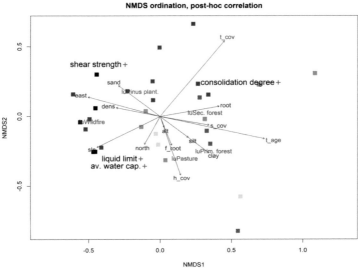

Fig. 2.6 (continued)

For the B Horizon *tree cover* and *tree age* are significant correlations of soil stability indicators ($p = 0.001$), but *vegetation assemblage* shows again the best correlation ($r^2 = 0.44$). With respect to the independent variables, it should be emphasized that secondary forests have significantly higher root penetration ($p = 0.001$) than plantations and even slightly higher than primary forests. Primary forests have significantly lower bulk density in the A Horizon than all other vegetation formations. Soil texture (as the most important property for soil stability) is consistently similarly unfavorable to the formation of stable properties in all plots.

The multivariate evaluation thus shows two things: First, there are indeed significant differences in soil stability between vegetation assemblages, especially if considered as a multidimensional characteristic. Second, these differences can be attributed to vegetation type or species composition and structural properties of the vegetation, while soil type and topography are negligible.

2.5 Discussion

In Sect. 5.1 each vegetation assemblage will be related to hillslope stability and specific plant-soil, or better said biogeomorphic interactions. The relevance of these interactions for biogeomorphic feedback systems will be established in Sect. 5.2. Subsequently, in Sect. 5.3, the risks and consequences of land use for landscape resilience (nature conservation and natural hazard mitigation) are outlined.

2.5.1 Vegetation Assemblage Specific Biogeomorphic Interactions and Hillslope Stability

2.5.1.1 Primary Nothofagus Forests

Primary *Nothofagus* forests have lower erosion rates than other land uses in Patagonia (Bonilla et al., 2014; Rostagno et al., 1999). Surprisingly, *shear strength* is found to be lowest there. The *consolidation degree* in primary forests is also just in the middle range of values. In contrast, *liquid limit* is significantly higher under primary forests than under plantations. This corresponds with the results of Candan and Broquen (2009) who found aggregate stability—which is an indicator similar as the *liquid limit*—under *Nothofagus* forests significantly higher than under *Pinus* plantations. Also, the *available water capacity* is highest under primary forests. Consequently, hydrological mechanisms and the maintenance of the water cycle balance are the characteristic biogeomorphic feedbacks (which will be described in Sect. 5.2) in the final successional stage.

2.5.1.2 Secondary Nothofagus pumilio Forests

Liquid limit of secondary *Nothofagus* forests shows a tendency towards the original values of primary forests, which is particularly relevant for the cohesion of andosols (Maeda et al., 1977). The *consolidation degree* of secondary forests is higher than those of other land uses. The reason is the root abundance of *Nothofagus pumilio*, especially in the B Horizon (Fig. 2.8). It is typically strongly horizontally interconnected and therefore withstands the strong winds in Patagonia (Sánchez-Jardón et al., 2010). Thus, in the case of disturbances of the vegetation, the stabilization of the soils is maintained for several years, bridging the period until new vegetation stabilizes the soil again (Korup et al., 2019). Thus, secondary forests create a positive, re-enforcing feedback loop. The "hillslope stability side" in particular is enhanced by the increased cohesion due to the higher *liquid limit*, which reduces the risk for rainfall-induced landslides and erosion. At the same time the roots stabilize deeper horizons. Vegetation fitness is strengthened via the growth characteristics of *Nothofagus pumilio*, which enable forest expansion even under unfavorable environmental conditions such as high solar radiation and wind speeds (Till-Bottraud et al., 2012).

2.5.1.3 Pinus Plantations

Pinus plantations have lower erosion rates than pastures (La Manna et al., 2016). At first sight, they fulfill their original implication. Nevertheless, the contribution of plantations to hillslope stability should be considered more closely. *Shear strength* is generally very low in all assessed soils, but significantly higher under Pine plantations. Within this group it is highest under *Pinus contorta* stands, especially in the A Horizon. One reason might be the slightly higher clay contents in the lower parts of the hillslopes with smaller slope angles, where those plots are located. Another reason might be the obvious extended Mycorrhiza (Fig. 2.4), which is characteristic for *Pinus* (Kimmins, 2003) and was probably introduced along with the *Pinus* trees (Barroetaveña et al., 2005). Sporadically plantations have a high *consolidation degree*. This is also due to mycorrhiza and not to the roots, as was the case with secondary forests. Therefore, the *consolidation degree* has to be evaluated differently, because the stabilization by mycorrhiza is a temporary effect, which can easily be lost, for example by changes in the soil air and soil water balance. (Scheffer et al., 2010, p. 209). This is evident from the Wildfire areas, where mycorrhiza has been lost and the *consolidation degree* is therefore in the most unstable category (Fig. 2.5).

The data show that *Pinus* plantation soils have the lowest *available water capacity* of all vegetation assemblages studied, with exception of the wildfire sites. At the same time, *Pinus* plantations have higher water consumption than native species forests and even than grasslands (Gyenge et al., 2003; Licata et al., 2008). Therefore, it can be assumed that plantations develop a higher risk of water shortage than other vegetation assemblages. On the one hand, this is relevant for hillslope stability as andosols lose their cohesion when they are too dry (Maeda et al., 1977). On the other hand this can increase the risk of wildfires, resulting in bare soils for a significant time period. During this period until new vegetation emerges, these soils are then subject to severe erosion and landslides.

Thus, although the measured soil stability indicators show partly good values, they should be evaluated differently as under secondary forests because the underlying processes—biogeomorphic interactions—have a different character. The vegetation effects of soil stabilization, both the hydrological and mechanical mechanisms, are quite vulnerable under plantations and jeopardize the biogeomorphic feedback loop. This concerns, on the one hand, the "hillslope stability side" of the feedback loop (right side in Fig. 2.1), but also, in particular, the "vegetation fitness side" (left side in Fig. 2.1): the increased wildfire risks are one aspect of the weakening of vegetation fitness, another one the negative impact on biodiversity. Under plantations, the herb layer, but especially the shrub layer is significantly reduced (Braun et al., 2017b).

2.5.1.4 Wildfire Sites (ex. Pinus Plantations)

The weak performance in all the assessed soil stability indicators on the Wildfire plots demonstrates that the hillslope stabilization and erosion protection by the *Pinus* plantations is a labile and temporary effect. In the case of a wildfire, the stabilizing properties are completely lost. During the field sampling campaign, which took place two months after the wildfire, massive erosion was already evident through several gullies at the hillslope and sediment load clogging drainage facilities in residential areas at the base of the slope.

2.5.1.5 Pasture

Pasture soils are compacted due to livestock (Fig. 2.8). In combination with the higher clay contents of the plots due to its location at the foot of the slope, this leads to the high values of *shear strength*. The compaction might also cause the high values in *consolidation degree* in the A Horizon. However, they have very unstable properties in the B horizon, among other things because there is hardly any root penetration. The low level of *consolidation degree* is probably the reason for the typical shallow landslides in the study area, where the weak layer is often

2.5 Discussion

located approximately in the depth of the B Horizon. The *liquid limit* is at very high levels, even significantly higher than under primary forests. The is probably due to the large amount of organic material that is brought in, whereby land use-related differences in the wetting-drying cycles can also have an influence on soil physical parameters (Dec et al., 2011).

Yet, in spite of the good values of *consolidation degree* and *liquid limit* in the A Horizon, the pasture sites are highly erosion-prone: Erosion experiments show that pasture sites are more susceptible to water erosion than soils under *Pinus ponderosa* (La Manna et al., 2016). Especially during rain events the soils are highly exposed to severe erosion due to the lack of interception (Rostagno et al., 1999). Without interception by woody species, the good values soil stability properties can nevertheless not prevent erosion and landslides (Parizek et al., 2002; Rostagno et al., 1999).

However, the Patagonian ones are nevertheless more resilient than Icelandic pasture soils, which are quite comparable in terms of soil parent material and prevailing climatic conditions. The reason for this are both hydrological and mechanical vegetation effects: the species found in Patagonian steppes are even better adapted to the climate, as they resist winds due to their growth forms and maintain the water cycle balance due to low evapotranspiration rates (Schmidt, 2022, pp. 155–157). However, without improvement of soil conditions and sheltering techniques after planting, woody species are unlikely to re-establish, which represents a disturbance of the vegetation fitness feedback (Valenzuela et al., 2018).

2.5.2 Impacts on Biogeomorphic Feedbacks

As described in the last section, different biogeomorphic interactions may underlie the soil-stabilization. These are critical to how biogeomorphic feedbacks are develop. In the long term, this will also determine whether landscape resilience is increased, maintain or even weakened. These relationships are portrayed and the main findings summarized in Fig. 2.7 and are discussed in the following.

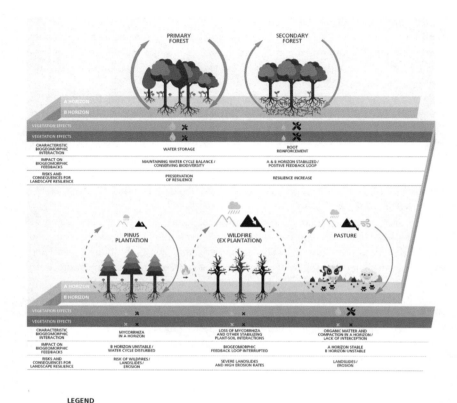

Fig. 2.7 Schematic overview of land use impacts on biogeomorphic feedbacks. The figure represents the impact of land use on hillslope stability, including mechanical and hydrological mechanisms, the respective causal plant-soil interaction (biogeomorphic interaction), the impact on biogeomorphic feedbacks, as well as the risks and consequences for landscape resilience. Size of symbols represents the quantity of the effect. Concept: Eichel, Tröger. Graphical design: Gesa Siebert (IfR)

2.5 Discussion

2.5.2.1 Mechanical Mechanisms

Among the mechanical mechanisms by which vegetation improves slope stability, two indicators were selected that express the stability of the soil: the *shear strength* and the *consolidation degree*. The *shear strength* is equally weak among all land uses and in absolute values, the differences in shear strength between land uses are marginal. Also, the macro stability of rooted soils is difficult to capture with this parameter. Therefore, the *consolidation degree* is of greater importance, which actually shows differences between the land uses. Since the soil texture is homogeneous, the differences result from biogeomorphic interactions. In turn, this stabilization is either by roots, as in the case of secondary forests, or by mycorrhiza, as in the case of *Pinus* plantations. Both represent biogeomorphic interactions. The same measured effect can thus have different biogeomorphic interactions as its cause. This in turn determines how the biogeomorphic feedbacks might develop: In case the mycorrhiza is lost, stability is abruptly gone, as evident via the Wildfires plots. The most stable biogeomorphic interaction leading to a high *consolidation degree* is stabilization by roots, which is significantly higher under secondary forests than under plantations ($p < 0.001$, Sect. 5.1, Fig. 2.9). The high levels of *consolidation degree* through compaction by livestock on the pastures, on the other hand, cannot prevent high erosion rates as interception by woody species is missing (a lack of biogeomorphic interaction). Other biogeomorphic interactions, especially biomechanical effects such as tree-uprooting (Pawlik, 2013), are omitted from the present study, but are also likely to contribute to slope stabilization in primary and secondary forests.

2.5.2.2 Hydrological Mechanisms

The *liquid limit* is related to mechanical and hydrological mechanisms and especially relevant for Andosols, as they can be saturated for a short time without losing their structure as long as there is no additional mechanical disturbance (Basile et al., 2003). *Liquid limit* clearly depends on the vegetation assemblage and land use, with secondary forests showing a tendency to develop similar values as primary forests. Pastures have even higher *liquid limits*, meaning that they can absorb more water before liquefying. However, pastures lack interception and mechanical stress from livestock might trigger soils to slide at water contents near the *liquid limit*. Therefore, the high values are only of limited use for hillslope stability of pastures, whereas secondary forests protect the soils from mechanical influences due to the above- and below-ground vegetation structure and therefore maintain hillslope stability even at lower values. This example particularly illustrates how soil properties and biogeomorphic interactions intertwine to produce a hillslope-stabilizing feedback.

Andosols have a high proportion of allophanes (Dörner et al., 2012). In contrast to crystalline clay minerals, their cohesion in the dry state is very low, making soils in the study area unstable at low moisture levels (Dec et al., 2012; Dörner et al., 2012; Maeda et al., 1977). Consequently, a higher *available water capacity* ensures that this occurs less likely and supports hillslope stability (Dörner et al., 2012; Dörner et al., 2009).

Although it has been shown in this study that land use has an influence on the water storage capacity of soils, questions remain regarding the water cycle balance and soil chemical processes. Different fractions of soil organic carbon (SOC) such as the labile fractions are related to the biological activity of the soil, as well as the passive pools are related to storage of carbon for long periods of time. These pools integrate structural and functional properties of the soil (Christensen, 1996; Zagal et al., 2002). Land use can generate changes in the SOC, e.g. the decrease of the contents and changes in the type of SOC, which can generate loss of the structural stability of the soil and thus favor runoff and erosion (Dec et al., 2011; Dörner et al., 2009). The function of the pores can also be impacted and with it the capacity of water storage in the soil (Dec et al., 2011; Dörner et al., 2012). The differences in soil properties observed in the present study might be explained to a large extent by the organic carbon input characteristic of each land use, especially the hydrophobic properties of the needle litter of *Pinus* trees. Hence, future evaluations could include further soil indicators and properties such as hydrophobicity, bulk density, porosity and so on.

The *available water capacity* also has relevance for the other side of the biogemorphic feedback loop, the vegetation fitness: *Nothofagus pumilio* seedling survival benefits from improving soil properties before planting, especially those having an effect on water availability (Valenzuela et al., 2018). This would represent a biogeomorphic manipulation that strengthens the feedback loop. The vegetation fitness can also be impacted by wildfire risks resulting from low *available water capacity*, as discussed in Sect. 5.3.

2.5.3 Land Use, Biogeomorphic Feedbacks, Risks and Consequences for Landscape Resilience

Besides its role in hydrological mechanisms of hillslope stabilization, *available water capacity* is relevant to landscape resilience via influencing the risk of wildfires. The severity of wildfires is influenced by soil moisture (Krueger et al., 2015). As *available water capacity* was shown to be depended on vegetation assemblage, with soils of *Pinus* plantations having lowest capacities, it can be assumed that the soil water content varies accordingly depending on land use. Besides having the lowest capacities, *Pinus* plantations additionally have a higher water consumption than other vegetation assemblages in Patagonia (Gyenge et al., 2003; Licata et al., 2008). From the combination of these arguments, it can be concluded that the risk of wildfires is increased in *Pinus* plantations compared to secondary native forests, and additionally would leave soils more susceptible to erosion and landslides: the *consolidation degree* the *Pinus* plantations provide, is based on mycorrhiza stabilization and is lost with it in the case of wildfire. In contrast, *consolidation degree* in secondary forests is based on roots also in deeper soil horizons, stabilizing soils for several years even after disturbances of above ground vegetation (Korup et al., 2019). Furthermore, *Nothofagus* forests are in general not very susceptible to wildfires (Paritsis et al., 2015).

The risk of wildfires depends not only on the vegetation assemblage, but additionally on land use management. Wildfire risks could be decreased via reducing tree density, branching measures or laying of firebreaks (Tedim et al., 2016). An adapted vegetation structure and composition could have further slope-stabilizing effects: different height levels, especially the existence of a relevant shrub layer with diversity in growth height reduces the kinetic energy of raindrop throughfall and thus mitigates erosion by the splash effect (Senn et al., 2020). In the shrub layer, the species composition also has an influence on forest fire risks, since the different species in Patagonia have different degrees of flammability (Blackhall and Raffaele, 2019).

2.6 Conclusions

A biogeomorphic approach was adopted to investigate if forest plantations are suitable for meeting their conservation objectives of erosion protection and landslide mitigation. Using soil indicators, it was shown that soil stability properties are dependent on vegetation assemblage. Primary forests base their landscape resilience on biogeomorphic feedbacks, with maintaining the water cycle balance being the most characteristic effect for hillslope stability and biodiversity for vegetation fitness. Secondary forests improve hillslope stability mainly through mechanical effects. Crucial to this is the high *liquid limit* in the A Horizon and the interconnected root system in the B Horizons. Thus, the biogeomorphic feedback loop is enhanced. *Pinus* plantations cause a weak improvement in soil properties relevant for hillslope stability, but with trade-offs in water balance and vegetation fitness. They hardly allow for a establishing a shrub layer and native species are reduced compared to secondary forests. Landscape resilience is thus impaired, especially by the higher risk of wildfires, which subsequently leads to severe erosion and landslides. Pastures show good values in the soil stability parameters, but their biogeomorphic feedbacks are too weak to rebuild landscape resilience and carry the risk of erosion and landslides.

The biogeomorphic framework allows the integration of different environmental aspects and therefore synthesizing the consequences of land use for nature conservation and natural hazard prevention. It might be useful to communicate results and insights about the impact of land use on environmental risks, e.g. for land use planning or other multidisciplinary communications and discourses. It is useful to sort ideas and findings in three levels: (I) the biogeomorphic interaction itself (where soil indicators were assessed and soil processes were shown to be vegetation assemblage and land use specific), (II) the possible impacts on biogeomorphic feedbacks, considering both sides of the feedback loop, and (III) the consequences for landscape resilience (Fig. 2.7).

2.7 Annex

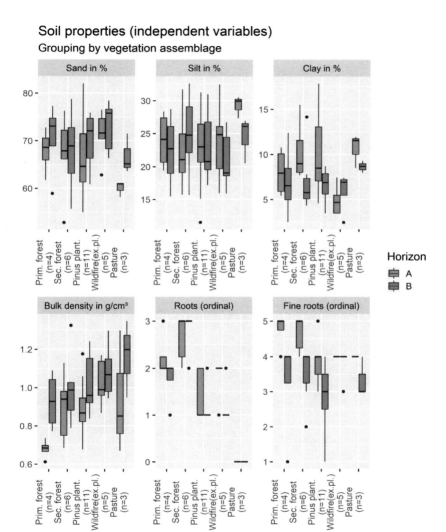

Fig. 2.8 Boxplots of the soil properties (independent variables)

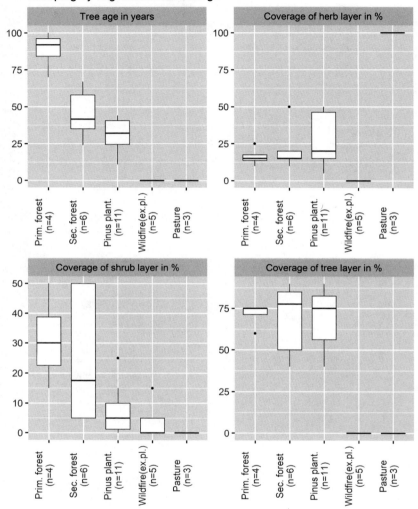

Fig. 2.9 Boxplots of the vegetation properties

"Industry Impacts More Than Nature": Risk Perception of Natural Hazards in More-than-human Worlds

3

This chapter is a reprint of the following article:
Tröger, Danny; Braun, Andreas Christian (2024): "Industry Impacts More Than Nature"—Risk Perception of Natural Hazards in More-than-human-worlds. In: International Journal of Disaster Risk Reduction, S. 104568. https://doi.org/10.1016/j.ijdrr.2024.104568.

Abstract

Understanding the risk perceptions of natural hazards is vital for disaster mitigation. Despite ex- tensive research on the topic, local and indigenous perspectives often remain marginalized and underrepresented. One reason discussed is the inherent Eurocentrism in knowledge production. Using narrative interviews with fishermen in the Patagonian fjord lands, we developed a grounded theory on risk perception against the background of actor–network theory to adress this criticism.

The emergent order of risk perception research can be described by constructions of natural event systems, attributions of cause and blame, psychological processing and emotional strategies, terri- torial and professional identity, and perceptions of the national state. The distinction in risk science between natural hazards and environmental degradation, which represents the manifestation of the nature-culture divide, is not necessarily made by people.

By allowing ambiguity in the ontology of nature, including networks of human and non- human actors, a hybrid understanding can be developed. We propose the following basic concepts of this understanding: environmental transformation, human-environment relations, appropriation of nature

© The Author(s), under exclusive license to Springer Fachmedien Wiesbaden GmbH, part of Springer Nature 2024
D. Tröger, *Assessing Landscape Resilience*,
https://doi.org/10.1007/978-3-658-47274-0_3

and ownership structure, displacement of sustainable traditions, and regional common-pool re- sources. Classifying these perspectives as pre-modern corresponds to the separation of different cultures described by Latour and forms the basis for asymmetrical relationships, reproducing the hegemony of Eurocentric and anthropocentric knowledge production. The separation into hazard and vulnerability can be seen as a manifestation of the second separation and leads to the net- works of human and non-human actors that promote resilience being overlooked.

3.1 Introduction

Understanding the risk perception of natural hazards is crucial for reducing disaster risks, for instance, in the context of land-use governance (Bohle, 2008; Greiving, 2011; Oliver-Smith, 2013; Renn and Klinke, 2016; Sapountzaki et al., 2011; Turner et al., 2003; Wachinger et al., 2013; Wisner et al., 2010). Risk communication should occur at an equal level to increase the acceptance of risk reduction measures or, ideally, co-create them with affected communities (Aguiar et al., 2021; Bendito, 2020; Braun, 2021a; Schneiderbauer et al., 2021). One unresolved problem is the coloniality of knowledge production, which causes asymmetrical relationships, especially in the Global South (Arora and Stirling, 2023; Gaillard, 2019a; Ghosh et al., 2021; Peters, 2021; Ramsey et al., 2019; van Riet, 2021).

The perspective on natural disasters has changed significantly over the course of human history. Historically, catastrophic events have been ascribed metaphysical significance, such as the punishment of God. This understanding fundamentally changed—for some parts of global society at least—when natural events were "discovered" to be independent of human actions and objectively investigable. This understanding laid down the epistemic, technical, and organizational foundations for transforming and dominating nature to an unprecedented extent (Banerjee and Arjaliès, 2021; Polo Blanco and Piñeiro Aguiar, 2019). With industrialization and increasing opportunities, the focus on natural hazard mitigation has shifted towards technical hazard prevention, such as the construction of dams against floods(Pohl, 2008). In parallel, top-down disaster management has been developed in many nations according to their military organization models (Dombrowsky, 2008).

Today, there is a broad consensus in Disaster Risk Research that the disaster risk of natural hazards is the intersection of hazards with vulnerability, meaning that natural disasters are actually natural event systems and social disasters (Birkmann et al., 2016; Birkmann et al., 2013; Birkmann, 2007; Felgentreff and Glade, 2008; Greiving, 2002; Hossain et al., 2020; IPCC, 2022; Wisner et al., 2010). In climate change science, risk also encompasses the dimensions of societal response and social-ecological resilience (IPCC, 2022).

The differentiation of scientific disciplines and the development of corresponding expert knowledge regarding hazards were accompanied by an increasing discrepancy between expert analysis and laypersons' perspectives (Renn and Klinke, 2016; Slovic, 1987). This discrepancy brought about the emergence of risk perception research as a field to analyze and describe this discrepancy (Slovic et al., 2011). However, combining different knowledge systems concerning risks is a challenge that persists today (Ramsey et al., 2019). Moreover, the generation of knowledge on this topic has been mostly Eurocentric, reproducing coloniality through epistemic violence (Katanha and Simatele, 2019; Selwyn, 2023; van Riet, 2021).

To overcome the gap between different knowledge systems and decolonize the research field of risk perception, we aimed to establish an approach grounded in actor–network theory (Katanha and Simatele, 2019; Ramsey et al., 2019). The approach removes asymmetry by describing risk perception with a nature and environment ontology based on those affected while simultaneously making the conditions under which knowledge arises explicit or deconstructing those of previous approaches (Haraway, 1988). This approach focuses on comprehensibility, making the obtained results useful for risk communication, risk governance, and further trans- and interdisciplinary research.

3.2 State of the Art Theoretical Design

The theoretical discourses we refer to include theories and approaches on natural hazard risks and their perceptions, as well as actor–network theory. The latter turned out to be helpful for the critical epistemic reflection of disaster risk research, and we explored its applicability to research on risk perception.

3.2.1 Conceptualizations of Risk and its Perception

Although we are focusing on risk perception in this study, it is also relevant to reflect on the epistemic history of research on natural hazards and risks, as it is intertwined with what academic knowledge production has constructed regarding laypeople's perceptions.

In natural sciences, engineering, and economics, risk is usually defined as the product of the extent of damage and the probability of its occurrence (Renn et al., 2007). This perspective, which is regarded as "objective," is challenged in particular by feminist perspectives, which pay much more attention to the contexts of knowledge production (Haraway, 1989; Olofsson et al., 2016). The *Riskscape* concept by Müller-Mahn et al. (2018) combines the physical-material side of risk with the discursive and constructivist dimensions of risk, that is, how people perceive, communicate, and manage risk. Within this concept, conceptual dimensions such as temporality, subjectivity, social group, plurality, spatiality, social practice, and power relations play crucial roles in shaping our understanding of risk (Müller-Mahn et al., 2018). This riskscape concept, which is neither purely realist nor purely constructivist, places particular emphasis on the relationships between the routines and everyday practices of people and societies that shape risk.

According to the *Disaster Pressure and Release* model, risk is defined as the intersection between exposure to hazards and vulnerability (Wisner et al., 2010, pp. 49–62). Vulnerability can be physical, ecological, social, economic, cultural, or institutional in nature (Birkmann et al., 2013). The progression of vulnerability is explained as follows (Wisner et al., 2010, p. 52): the *root causes* of vulnerability are political and economic-ideological conditions, limited access to power and resources, and a lack of state institutions. *Pressure factors* such as a lack of local investment, deforestation, and rapid urbanization transform these into *unsafe conditions* for certain social groups. The proximity of these three stages of progression to a risky situation can be spatial, temporal, sociogenetic, or a combination of these. In a recent extension of the model, structural violence was included as another group of drivers of vulnerability, including the historical patterns of "underdevelopment," colonial history, neo-colonialism, and neo-liberalism (Hammer et al., 2019).

The counterpart to the vulnerability of social groups is their resistance to natural hazards and their coping capacity, referred to as *resilience* (Turner et al., 2003). Disaster resilience is also considered a pillar of sustainability and, in some countries, one of the tasks of spatial planning (Berkes et al., 2006; Greiving, 2002). In the context of climate change impact research, risk is understood as the intersection of hazards, vulnerability, exposure, and responses (IPCC, 2022, p. 147).

Lay people perceive risks differently than experts. Judgments under uncertainty follow certain psychological patterns, referred to as cognitive heuristics, many of which are relevant to risk perception(Kahneman and Tversky, 1974). One example is the affect heuristic: the feelings associated with potential sources of danger determine the judgment as a risk more strongly than "rational" considerations (Slovic et al., 2007). Generally, people evaluate risk based on qualitative characteristics, which is known as the psychometric paradigm. The multitude of criteria can be condensed using principal component analysis into two main characteristics: (un)familiarity with the source of risk and severity of the possible consequences. One of the underlying criteria relevant to the risk perception of natural hazards is the origin of the risk source (Slovic, 1987; Slovic et al., 1984). In general, a higher risk is accepted for natural hazards than for technical hazards, and paradoxically, experience with natural disasters often leads to lower awareness and less adaptation (Wachinger et al., 2013). Qualitative criteria of risk can be aggregated into semantic patterns of risk perception. For example, risk can be perceived as a stroke of fate, an early indicator of emerging danger, a challenge to one's strengths, or an immediate threat (Renn et al., 2007, p. 80). Apart from these effects, people in general, often act contrary to their conscious thought-out decisions, wherein cognitive dissonance occurs, and the attitude is subsequently reinterpreted (Festinger, 2001).

The attitude of reference persons or groups, the media landscape, and stakeholders, among others, can lead to a *social amplification of risk,* which means that the social environment can reinforce feelings of insecurity, even if a particular person was initially unafraid of the risk and was not directly affected themselves (Kasperson et al., 1988). The perception of natural risks and their management is influenced not only by the actual social environment but also by the desired organization of a society, including cultural group identity. Furthermore, this influences the characteristics that are attributed to nature, for example, whether it is stable or unstable, and the connections between the idea of the optimal level of organization of society, nature myths, and the associated acceptance of natural hazards (Douglas and Wildavsky, 1983).

In the risk management framework of Renn et al., risk perception is one of eight criteria used for classifying risks, leading to different management strategies (Klinke and Renn, 1999; Renn et al., 2007). Risk Perception is relevant for the mobilization potential criterion, that is, the extent to which civil society demands arise in connection with risks. It contains four aspects: 1) psychological stress with regard to the risk and source of risk, 2) possible spillover effects, i.e., negative consequences in other socially relevant areas, such as the economy, 3) potential for social conflict and mobilization, and 4) inequality and injustice regarding the distribution of risks and benefits across space, time, and social status. The perception of vulnerability was not included.

The phenomenon of risk perception (each vast field of research) can also be grouped into Heuristics of Information Processing, Cognitive and Affective

Factors, Social and Political Institutions, and Cultural Background (Renn and Klinke, 2016). Although the outlined approaches are useful, they have limitations, especially in the context of the Global South (Ramsey et al., 2019; van Riet, 2021).

3.2.2 Recent Critics and Potentials of the Actor-Network-Theory in Disaster Risk (Perception) Research

A recent point of critique in the disaster risk reduction community is that "studies often overlook or oversimplify the understanding of risk perception of indigenous communities. (...) Neocolonial patterns, including a colonial and neocolonial understanding of the environment, a hierarchy of knowledge, and a reproduction of Eurocentrism, have marginalized local people who have had little involvement in decision-making, even if those decisions affect their daily lives" (Schneiderbauer et al., 2021). Local ontologies and epistemologies, the capacities of local communities, and the root causes of vulnerability remain underscored (Gaillard, 2019a). The epistemology of risk research is currently being discussed as the cause of these shortcomings. These fundamental misunderstandings might be the consequence of the illusory simplification that there is a "human" and a "natural" world (van Riet, 2021). This idea of the so-called nature-culture-distinction in modern society originated from science and technology studies (STS) and Actor-Network Theory (ANT) (Haraway, 1989; Latour, 1995).

The main characteristic of modern society is that it determines the boundaries of nature and culture for almost any topic and discourse, which is a phenomenon called purification work (Latour, 1995). One example of a risk study is the definition of risk as the intersection of hazard and vulnerability (Birkmann et al., 2013). Another example is the distinction between natural processes that pose hazards and anthropogenic influences or natural hazards and environmental degradation. An enormous effort is undertaken to demonstrate by means of scientific knowledge production the "parts" of natural hazards are "natural" and those that are "anthropogenically influenced," thus being assigned to "culture" (Gill and Malamud, 2017). This separation between nature and culture is considered an internal separation in modern societies.

Based on this internal separation of nature and culture, Latour described a second, external separation, namely into modern and pre-modern "cultures." From the perspective of the "moderns," "pre-modern" cultures do not accomplish the internal separation (Latour, 1995), which results in a fundamental misunderstanding and asymmetrical relationships.

3.2 State of the Art Theoretical Design

ANT proposes ways to overcome this misunderstanding (Latour, 1995). Instead of discussing which parts of an object (quasi-object) are natural and cultural, the analysis should be based on which human actions are enabled or stimulated. Non-living entities might trigger human action or non-action, thus embedding social relations or "inscriptions" (Latour, 1995; van Riet, 2021), which have their own agendas, so to speak. This leads to human and non-human actors forming assemblages of more-than-human worlds (Anderson and McFarlane, 2011; Neisser, 2014).

Based on ANT, Latour developed a description and characterization of formal knowledge production, which he called the mode of existence Reference [REF] (Latour, 2014). It describes knowledge as a chain of references to newly generated entities, facilitated by technology and machines, which convey essential knowledge ("immutable mobiles") despite or precisely because of their dissimilarity with the object of investigation. Therefore, in the context of ANT, one could say that the risk and risk perception approaches outlined in the previous section are reference points for the mode of existence [REF]. One of the most important reference points for disaster risk research is the distinction between natural and anthropogenic hazards.

A comprehensive synthesis and outlook of the advances in analyzing disaster risk through the lens of ANT was presented by Neisser (2014), who emphasized the interrelations of actors and the conditions, mechanisms, or processes that constitute risk, thereby bridging realist and constructivist approaches (Müller-Mahn et al., 2018; Neisser, 2014). Using ANT, it becomes possible to bridge epistemic challenges between various scientific disciplines, overcome heterogeneities, accept hybridity, and avoid dichotomies and a priori distinctions, such as those between human/non-human or nature/culture. Thus, it is possible to include not only material and social entities, but also discursive entities, including competing understandings of risk (Neisser, 2014).

ANT has been applied to a handful of empirical disaster risk studies, such as eco-based natural hazard mitigation in the semi-arid regions of Zimbabwe (Katanha and Simatele, (2019), impacts of bioterrorist scenarios of anthrax letters (Maintz, (2008), and authoritarian Chilean government's response to the 1985 San Antonio earthquake as a cover for neoliberal reforms González (2022). However, ANT has not yet been applied to risk perception or research. Therefore, we aimed to make an exploratory attempt to do so in this empirical study. We begin with the hypothesis that one reason for the frequent underestimation and disregard of local risk perceptions is rooted in the epistemology of risk perception research. Although not sufficient on its own, we consider that abolishing this asymmetry is necessary to decolonize this field of research, introduce indigenous and other deviant perspectives, and dismantle further hegemonic power relations in the field of disaster risk prevention, thereby addressing epistemic violence.

3.3 Study Site

The study area is the Aysén fjord (45°21'36" S, 73°04'48" W), which is located in the Chilean-Patagonian Fjordlands. It is the main access point for the region, as there is no continuous road connection between the Chilean territory and rest of the national territory.

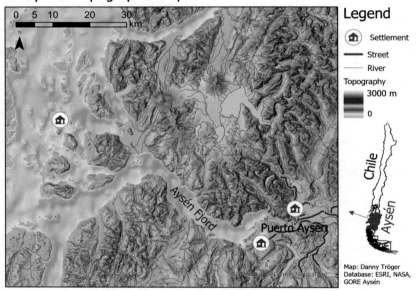

Fig. 3.1 Topographic Map of the Study Site. (Cartography: Danny Tröger, data: ESRI, NASA, Regional Government Aysén)

3.3.1 Physical Geography

The study area (Fig. 3.1) lies to the west of the main Andean Ridge, which has a maximum height of approximately 2,600 m above sea level at these latitudes. It is fully exposed to a planetary westwind zone from the Pacific Ocean. These fjords cross the volcanic arc and fore-arc basins of the South American subduction zone. Most valleys run east-west, but some also run north-north-west along the Liquiñe-Ofqui Transform Fault Zone (LOFZ) (Vargas et al., 2013). The climate is characterized by harsh humid winds, with rainfall up to 3,000 mm•a^{-1}, temperatures between 2 and 18°C and frequent wind speeds of 9–12 m•s^{-1}. It is classified as type CfB in the Köppen/Geiger effective climate classification. Glacial retreat after the ice ages occurred more slowly than in Europe owing to the cold Humboldt Current (Glasser et al., 2008). Therefore, these fjords are younger than those in Norway, and some glaciers calve directly into the sea (Glasser et al., 2008; Rivera and Bown, 2013). In many areas along this ridge, thick glacial sedimentary layers with thin, humus-rich soils lie on top of igneous rocks. The natural vegetation consists of temperate evergreen forests, with the characteristic species being *Nothofagus dombeyi (Oberdorfer, 1960)*. Coastal primary forests that have never been cleared are present in the west of this area, which contrasts the eastern parts of the transition zone from Andean forests to the steppe, where vast areas have been cleared for extensive livestock and agriculture (Braun et al., 2017a; Schmithüsen, 1956; Tröger et al., 2022). These natural conditions give rise to a series of natural hazards in the study area (Fig. 3.1 and Fig. 3.2).

3.3.2 Natural Hazards and Environmental Degradation

The LOFZ transform fault zone and volcanic activity have caused near-surface earthquakes (Cembrano and Lara, 2009; Vargas et al., 2013). A special phenomenon in this area is the occurrence of seismic swarms. This phenomenon involves a series of almost daily earthquakes that can occur over several months. In the last 100 years, this phenomenon has occurred at least three times in this region (Naranjo et al., 2009; Sepúlveda et al., 2010).

Volcanoes in the study area pose various hazards, such as explosive eruptions that release large amounts of ash or tephra fall and pyroclastic flows (Amigo, 2013; Wilson et al., 2012; Wilson et al., 2011). Ash layers on hillslopes can cause lahars, which can be triggered by melting glaciers and snow cover, covering distances up to 40 km (Amigo, 2013).

The intense relief dynamics in Patagonia cause additional types of landslides, such as glacial, fluvial, seismic, and volcanic. In addition, there are special phenomena, such as glacial lake outburst floods (GLOFs) (Anacona et al., 2015). Landslides in the Aysén Region can even become maritime hazards: rock falls in the fjords can trigger an event locally referred to as 'local tsunamis', more accurately known as flood waves, of up to 40 m in height at the rockslides opposing the fjord flanks and 10 m even at a distance of 7 km from the original landslide (Naranjo et al., 2009; Sepúlveda et al., 2010). Using bathymetric surveys, we found that at least four such events occurred in the Aysen Fjord during the Holocene (van Daele et al., 2013).

Coastal ecosystems are under pressure from environmental degradation from aquaculture and tree plantations. Industrial aquaculture (salmon farming) affects water quality by increasing nutrients, leading to eutrophication, drug residues, parasites, diseases, and changes in the population dynamics of predatory fish and sea lions (Barton and Román, 2016; Buschmann et al., 2009; Iizuka and Zanlungo, 2016). In the catchment area of the Aysen River, rapid deforestation by slash-and-burn resulted in increased sediment, causing siltation of the Puerto Aysén harbor and changing flood-prone areas (Armesto et al., 2010). Tree plantations with exotic species have been planted around Coyhaique to counteract erosion (Braun et al., 2017a; Tröger et al., 2022). Tree plantations have been shown to affect marine biology in south-central Chile, leading to lower revenues for small-scale fishermen and forcing them to extend their routes (van Holt et al., 2017; van Holt et al., 2012).

3.3.3 Human Geography and Socio-Environmental Conflicts

Before colonization by the Criollo people and European settlers, the study area was inhabited by the indigenous people of Chonos and Kawésqar, who lived on the coasts as half-nomads. More indigenous communities arrived throughout the last two centuries, and many of the Mapuche were forcibly expelled from south-central Chile at the end of the 19th century and fled to the Aysén region (Carruthers and Rodriguez, 2009b; Delamaza et al., 2017). Several coastal fishery communities and subsistence farmers in the eastern region descended from these groups. From the beginning of the 20th century, the Chilean state moved ahead with the permanent settlement of the region. Ranches and towns were primarily established in the eastern part of the region, where the land was made usable for

3.3 Study Site

agriculture and cattle breeding through slash-and-burn agriculture (Bizama et al., 2011).

In the 1950s, industrial fishing established itself as an economic pillar in Puerto Aguirre, an island off the Aysén Fjord, by constructing a processing factory (Molinet et al., 2014; Ovalle, 2011). In 1974, the year after the coup by Pinochet, coastal waters were declared the responsibility of the Ministry of Defense (D.S. 475 de 1974, Ministerio de Defensa Nacional), thus laying the legal foundation for using coastal resources, which established itself as an economic actor in Chile since the late 1980s, and in the Aysén region since the 1990s (Molinet et al., 2014). Owing to the long transport routes, only a few companies initially settled in the region. However, this changed abruptly in mid-2007 with the outbreak of the ISA epidemic in salmon farming in the 10th region, causing companies to relocate their production as an evasion strategy (Iizuka and Zanlungo, 2016).

In the Aysén region, repeated conflicts over environmental and developmental issues have occurred over the past 30 years, some of which have been accompanied by violent protests (Barton and Fløysand, 2010). One of the main environmental discourses regarding the tension between field development and conservation revolves around the salmon industry (Bachmann-Vargas and van Koppen, 2020). Although salmon farming is a large-scale enterprise, the economic yield in the Aysén region is relatively low because of weak supply chains and a lack of infrastructure (Bachmann-Vargas and van Koppen, 2020). Moreover, the low enforcement of environmental and conservation standards has led to opposition against aquaculture, especially from indigenous communities who are currently involved in legal disputes over maritime space (Bachmann-Vargas and van Koppen, 2020; Salgado et al., 2015).

3.3.4 Sociological Lense: Vulnerabilities

Fjords in the Aysén region are exposed to landslide hazards and flood waves. Various communities, several being indigenous, live informally along the coast and use the fjords as fishing grounds, although only a few of them can be found on official maps. These groups are referred to as traditional or small-scale fishers, locally called "pescadores artesanales." From a risk sociology perspective, communities are vulnerable in various ways, for example, due to a lack of existing infrastructure, low economic resources, institutional marginalization, and dependence on intact ecosystems and, most importantly, their boats. Waves in fjords caused by local tsunamis can easily destroy boats even when they are at great distances from the landslides. In addition, the fish stocks on which they depend are threatened by aquaculture. Salmon farms promote algal blooms or parasites that threaten fish stocks that local communities depend on (Barton and Fløysand, 2010; Buschmann et al., 2009; Iizuka and Zanlungo, 2016).

The second group exposed to natural hazards comprises workers in the aquaculture facilities of industrial salmon farming. These farms consist of several pontoons in which approximately 20 employees and workers live during shifts of usually 10–15 days. Hereafter, this group of people is referred to as industrial fishermen, locally called "unksalmoneros." They are directly at risk during their shifts, as pontoons can easily be destroyed by local tsunami waves while also being economically vulnerable. Salmon farms are only profitable if they are operated optimally and are quickly abandoned, and employees are dismissed if any difficulties arise, such as the spread of the ISA virus or the destruction of salmon farms by flood waves (Barton and Fløysand, 2010; Iizuka et al., 2016).

3.3.5 Disaster Event in 2007

In the Aysén Fjord, hundreds of landslides were triggered in the course of a seismic swarm, causing approximately 7,000 earthquakes that lasted from January to April 2007 (Sepúlveda et al., 2010). On 23 January (5.2 Mw) and 01 April (5.4 Mw), installations of the salmon industry were damaged by minor landslides. On April 21, 2007, an earthquake (6.2 Mw) caused several landslides that were sufficiently large to cause floodwaves in the Aysén Fjord (Naranjo et al., 2009), killing 10 people. Moreover, several salmon-farming facilities, aquaculture infrastructure, and houses were destroyed, and countless boats were damaged (Naranjo et al., 2009).

In the wake of the seismic swarm in 2007, a general political and social crisis emerged in the region (Soule, 2012). From a risk research perspective, two social factors that led to the crisis are worth highlighting. Experts have extensively discussed the type, genesis, and possible consequences of seismic activity (Marin et al., 2020). This scientific controversy added to the uncertainty among the population, who were already confused by the sensationalist reporting style of the local media and political instrumentalization by influential individuals (Marin et al., 2020; Soule, 2012). Second, there were massive administrative and institutional weaknesses, including lacking information on the locations of the salmon farms and emergency evacuation possibilities, unclear responsibilities paired with centralized, segmented state administration, and a lack of search and rescue (SAR) facilities, staff, and health infrastructure (Marin et al., 2020; Soule, 2012).

3.4 Materials and Methods

Our methodological approach is two-pronged: creating a risk map and constructing a grounded theory on risk perception based on interviews. The risk map illustrates the epistemic process by which academic knowledge about natural hazards is usually generated, which represents the benchmark against which laypersons' risk perceptions are evaluated.

3.4.1 Risk Map

The spatial risk analysis was based on the framework of Taubenböck et al. (2008) and intersects the hazard map with vulnerable elements, resulting in a risk map. The hazard layers are based on the datasets of the regional government and vulnerable elements (salmon farms from Salmon ChileAG). There is no representation of approximately 2,990 traditional fishermen in the region (Molinet et al., 2014).

The entire LOFZ transform fault was considered a potential source of seismic activity, and a 40-kilometer-wide buffer was placed around it using ArcGIS. All salmon farms located at distances greater than 40 km were removed from the dataset. This distance is derived from the events in 2007 and corresponds to the distances at which landslides were triggered in a magnitude 5.5 earthquake (Keefer, 1984; Oppikofer et al., 2012). Subsequently, the distance of the salmon farms from potential landslide areas was divided into three classes (0–1 km, 1–4 km, and 4–7 km). The class boundaries were determined based on the impact of the 2007 disaster (Naranjo et al., 2009).

3.4.2 Risk Perception

Our approach to understanding risk perceptions was constructed in two stages. As a basis for grounded theory, the interviews were first analyzed using the reconstructive method (Bohnsack, 2014).

3.4.2.1 Selection of Interview Partners and Size of the Data Set

Field research was conducted from April to July 2016. Local gatekeepers were used to establish contacts. Notably, some interviewees approached the site spontaneously. Subsequently, further contact was established using existing methods to develop a snowball sampling system. We followed the principle of theoretical saturation to determine the number of interviews required (Draucker et al., 2007). This means that no further interviews will be conducted if a coherent theory is formed. A list of metadata from the 24 interviews can be found in the Appendix.

3.4.2.2 Interview Form and Strategy

3.4.2.2.1 Narrative Interviews and Focus Groups

We conducted narrative interviews and two focus groups with fishermen (Heinze, 2016; Schulz et al., 2012). At the beginning of the interviews, a narrative-generating question was asked. The respondent was asked to describe the important events and changes in their life over the past 10–15 years, which includes the 2007 seismic crisis. Following this, the respondents were asked questions by the interviewees about relevant topics raised (immanent questions) (Bohnsack, 2014). Subsequently, topics that were not mentioned by the respondents were asked by interviewers (exmanent questions). These topics followed a thematic guideline: first, what does the respondent think about the Aysén Fjord and the natural conditions, followed by which aspects they consider as challenges. Subsequently, respondents were asked about the impact and importance of natural hazards. If the respondent had not done so, the topic of landslides was initiated using the example of the 2007 crisis. The focus group interviews started immediately with several questions.

3.4.2.2.2 Expert Interviews

Expert interviews were conducted with actors relevant to industrial activities, risk mitigation, and spatial planning, whereby a specific guideline was designed according to the respective positions of the interviewees and adapted to the new insights during the interview.

3.4.2.3 Evaluation and Grounded Theory Construction

All interviews were transcribed and analyzed according to the documentary method, which consists of four steps: formulating and reflecting interpretation, case description, and case comparison (Bohnsack, 2014; Bohnsack and Nentwig-Gesemann, 2011). This interpretation divided the transcribed interviews into thematic paragraphs. The sections were then interpreted reflectively, considering the context, such as the interview situation, profession of the interviewee, and socioeconomic background. Theoretical, implicit, and tacit knowledge was reconstructed. The descriptions summarize the respective cases. General patterns and concepts were established through case comparisons.

Our original assumption was that determinants of habitus, such as occupation, religious affiliation, and ethnicity, could lead to different risk perception patterns, and that we could build a theory on the influence of those dimensions on risk perception. We also expected determinants such as origin and education to influence risk perception. However, it turned out that this is only partly true and quite narrow. No general patterns emerged; in particular, current identity hardly reflected the diversity of the trajectories we encountered. As expected, there were indigenous fishermen working in the industry who descended from European immigrants living semi-nomadically, and white managers related to the Pacha Mama, not to the Christian God. Therefore, we decided to construct a grounded theory (Breuer et al., 2019; Corbin, 2011; Diehl et al., 2022).

The following strategy was pursued to construct the grounded theory. The concepts were constructed in correspondence to the emergent order of risk research, as outlined in Sect. 2.1 (more details are outlined in the discussion). The term emergent order, inspired by social system theory, describes a meta-structure or meta-logic that emerges from the subparts of a system, but is more than the sum of its components.

Concepts that do not fit the premises of risk research are referred to as hybrid understanding. This means that they do not fulfill the following conditions: the distinction between natural hazards and environmental degradation, the linear concept of time with the distinction between disaster and risk based on it, stringent cause-effect chains, and the national state and formal institutions as the only valid form of collective organization for disaster risk management. Why did we do that? Conventional research on risk perception overlooks these aspects, skipping the interview sections. For example, when addressing the risk of local tsunamis, an interview response indicating that technologies against algal blooms are unfortunately too expensive would, at best, be evaluated as an aspect of vulnerability.

The terms for the concepts of emergent order are inspired by theories and approaches of risk research and hybrid understanding rather than by discourses of political ecology, actor-network theory, sustainability science, and postcolonial studies. They are not to be regarded as final and complete but, instead, as a starting point for further elaboration and initiation of discourse.

3.5 Results and Discussion

In the following, the results and discussion are not separated into main levels but according to the three subject areas of risk analysis, emergent order, and hybrid understanding.

3.5.1 Risk Analysis

The risk analysis is essentially based on geoscientific hazard maps, which are the most common in risk governance.

3.5.1.1 Risk Map

The risk map (Fig. 3.2) visualizes the exposure of salmon farms to the hazards of pyroclastic falls (volcanic outbreaks), landslides, and the resulting local flood waves (also called local tsunamis).

3.5 Results and Discussion

Risk Map Aysén Fjord

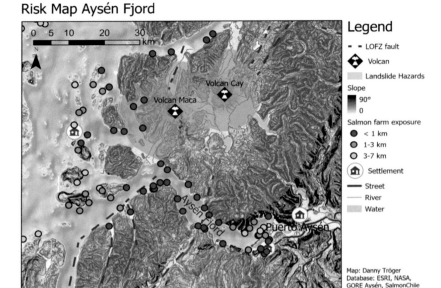

Fig. 3.2 Risk Map. (Method and cartography: Danny Tröger; data: ESRI, NASA, Regional Government Aysén, SalmonChile AG)

In the Aysén region, 132 of the 462 salmon farms are located <1 km from potential landslide areas and are, therefore, at high risk. Furthermore, 97 salmon farms were within 1–4 km, and 20 were within a distance between 4–7 km. The number of people exposed to the risk was calculated (Table 3.1) and differentiated according to the three exposure classes. Typically, a salmon farm is constantly run by 12–20 workers. This amounts to a maximum number of approximately 5,000 people being exposed to landslide-triggered tsunami hazards. Among these, approximately half work on pontoons <1 km potential landslide zones, which is approximately 2,600 people.

Table 3.1 Estimation of the number of persons at risk. (the calculation goes beyond the areas shown in the Risk map)

Exposure (distance in [km])	No. of salmon farms	No. of people (min.)	No. of people (max.)
0–1	132	1,584	2,640
1–4	97	1,164	1,940
4–7	20	240	400
∑	249	2,988	4,980

3.5.1.2 Reflections on Inherent Asymmetries

The risk map can be used to highlight the epistemic and ontological incentives between experts' and laypeople's perceptions of natural hazard risks that lead to asymmetric relations. The production of risk maps with an intersection of hazards and vulnerable entities is a standard procedure in risk governance and spatial planning instruments for disaster risk mitigation (Greiving, 2011; Greiving and Pohl, 2011; Taubenböck et al., 2008). They also form the basis of risk communication and participation practices (Sapountzaki et al., 2011). As Porter and Demeritt (2012) stated, "a risk or vulnerability map is a symbol, a discursive element, and a tool for political consulting or a mediator between scientists, stakeholders, and the public. Therefore, the map potentially functions as an obligatory passage point in the planning process". The intention is to provide an "objective" elicitation of hazards, aiming for their objective representation, which is problematic for epistemic and ontological reasons.

At the problem level of knowledge production (epistemic level), the inherent limitations of risk maps stem from the data and the methodology. The lack of sufficiently accurate input data for running landslide models is a rule rather than an exception in peripheral areas such as Patagonia. A closer look at the landslide hazard zones revealed that they were calculated using an indicator-based methodology, most likely based on remote sensing data. The problem is that decisions on category formation, such as thresholds for spatial resolution, slope angles, and probability of occurrence of landslides, usually have a greater impact on map accuracy than the improvement of the methods themselves (Braun, 2021c;

Robbins and Maddock, 2000; Rosa, 2018). Moreover, the landslide maps indicated landslide occurrence zones but not the extent and interactions with other hazards (Steger et al., 2021). However, data availability does not solve this problem; landslide risk mapping also continues to fail even in the Alps, a location with higher data availability and easily available experts (Hartmann et al., 2021). Additionally, even if the hazard maps were calculated exhaustively, vulnerable entities would still be barely representable. However, conscious and unconscious decisions about what is represented and what is not have techno-political consequences (Bennett et al., 2022; Loconto and Rajão, 2020; Rothe, 2017). In our study, fishing boats, informal settlements, and vulnerability and ecological threats to the resilience of small-scale fishermen were not represented, despite vulnerability and resilience having spatial patterns (Cutter et al., 2014; Holifield et al., 2010). Consequently, traditional fishers were barely recognized as a vulnerable group and are thus underrepresented in disaster risk reduction procedures.

The risk map provides an ontology of academic knowledge on natural hazards in an area/space. To provide practitioners of risk governance with a better understanding of the discrepancy in the risk perception of laypeople, their perceptions should be described by complex scientific models and approaches to risk perception (Poliseli et al., 2022; Rosa, 2018). However, the process remains entrenched in the structure where "real" hazards, risks, and vulnerabilities are defined a-priori, for instance, using a risk map (Weichselgartner and Pigeon, 2015). The risk map, as the representation of academic/scientific knowledge, would serve as the yardstick for measuring how risk perception works, so to speak (Astaburuaga et al., 2023). In our study, salmon farms were determined as the vulnerable elements exposed to natural hazards, indicating they were at risk. Thus, traditional fishermen were considered part of this risk. As discussed in more detail in Sect. 5.2, the distinction between natural hazards and environmental degradation, inherent in academic knowledge, is not adequate to describe the ontology of the risk perception of the affected communities.

To put it in the words of ANT and the modes of existence concept (Latour, 2014), research on risk perception was trapped in the reference chain of the existence mode [REF], for example, academic knowledge production. Previously, it was established in an earlier step that natural hazards and environmental degradation should be distinguished. Therefore, latter research on the perception of hazards was based on this distinction. According to Latour (1995, p. 127), this creates an asymmetrical structure: it is society itself that declares the "truth" in the present case about the space containing natural hazards and environmental degradation being a different subject. The perceptions of those affected were structurally excluded because the environmental impacts of salmon farms were not defined as natural hazards. Thus, perceptions of environmental impacts were excluded from studies on the risk perception of natural hazards. However, this can be interpreted as an example of the maintenance of nature-culture separation, which Latour calls "purification work." People who do not make this distinction (in the eyes of the "moderns") were consequently seen as "pre-modern," which forms the basis of their marginalization through epistemic violence (Sullivan, 2017).

3.5.2 Risk Perception

An overview of the grounded theory of risk perception is shown in Fig. 3.3. The *emergent order* sub-concepts are juxtaposed with those of *hybrid understanding*. The left-hand side relates to the established approaches and theories of risk perception (emergent order). The right side refers to a more-than-human vision inspired by the actor-network theory (Latour, 1995), with a different ontology of risk related to the perception of the environment.

3.5 Results and Discussion

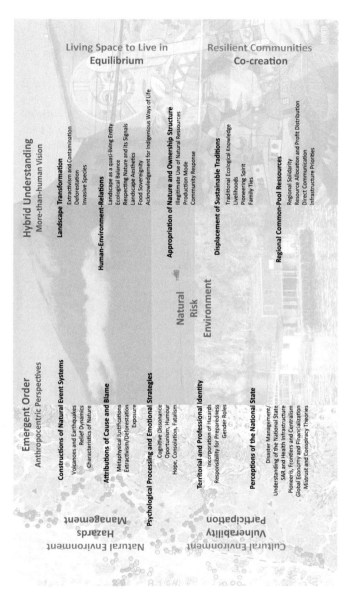

Fig. 3.3 Visualization of the grounded theory of risk perception. The emergent order on the left, and the hybrid understanding on the right side. The first one follows the distinction of natural hazards and environmental degradation, representing the manifestation of the nature-culture-distinction (First, internal distinction according to Latour). In contrast, the hybrid understanding reflects a more-than-human vision as we found it in narrative interviews. Interpreting these perspectives as "pre-modern"—corresponding to the second, external distinction according to Latour—would be the ontological foundation of asymmetric relationships, consequently reproducing coloniality and the hegemony of eurocentric knowledge production. (Map and photos in the background: Danny Tröger. The picture in the background on the right shows the façade of the "House of Cultures" in Puerto Aysén)

3.5.2.1 Emergent Order: Findings and Reflections

The *emergent order* of risk perception is structured into *constructions of natural event systems, attributions of cause and blame, psychological processing and emotional strategies, territorial and occupational identity,* and *perceptions of the nation-state.*

The concepts of *emergent order* cover the spectrum from the natural environment and its hazards to the cultural environment and vulnerability, which aligns with risk frameworks, such as the pressure and release models outlined in the introduction (Birkmann et al., 2016; Wisner et al., 2010). These are considered extensible archetypes, and the sub-concepts and concrete examples may vary depending on the respective hazard and study area.

3.5.2.1.1 Constructions of Natural Event Systems

Volcanoes and earthquakes. The Aysén region was considered to have low earthquake risk. Seismicity was attributed to volcanic activity, and we only found one case being attributed to plate tectonics. Volcanoes are undercut and felled by burning oil, eventually forming unstable caves that might collapse. Dust clouds from large landslides were interpreted as volcanic ash.

> "They are an accumulation of volcanic plates, and the intense burning of rock creates high pressure below. No matter how firm it is, the rock will have to give way and will have to break, resulting in an earthquake"

Relief dynamics. The mountains in the region were described as unstable because of their very thin vegetation cover. Heavy rainfall events have been recognized as triggers of landslides. Notably, they were not associated with the post-glacial processes present in the area during any interview.

> "because when it rains a lot, the, (...) trees become overloaded with water, causing landslides without earthquake"

Characteristics of Nature. Nature is fragile and can be easily disturbed by human activity. This attribution is found in numerous interview statements, and demonstrates the construction of the states of natural systems described by Douglas and Wildavsky (1983).

3.5 Results and Discussion

3.5.2.1.2 Attributions of Cause and Blame

Metaphysical justifications. In addition to the Christian interpretation that natural disasters were thought of as punishments from God for the sins of humans, other subjects referred to the Pacha Mama intentionally responding to human (mis)behavior with volcanic eruptions.

> *"Hebrews 13:21, how it says, (slow) for the sake of justice, (…), god destroyed the land"*

> *"or mother earth, the pacha mama, as they say here (…)"*

Resource Extraction. The extraction of petroleum is credited with destabilizing the earth.

> *"From my perspective, I think that the earthquakes, tremors, and so many other issues are effects… of oil extraction."*

Deforestation. Deforestation was classified as a serious environmental change.

> *"It has even been seen, eh, (laughing) the Ayseninos did not take long at all to make it firewood. There was a, (laughing) heh heh.. at the exit of the fjord—that is, there was a coigue that was about 2 meters in diameter along the Simpson River, ..yes, it was huge, and.. (laughing) they did not take long at all to make it firewood—with a chainsaw and axe they chipped it all down in a week"*

Exposure. Chile was assumed to be exposed to all types of natural hazards.
SC: *"Chile is a country super exposed to this type of phenomena."* N: *"yes,… exactly, earthquakes, tsunamis, (laughing) everything.. are everywhere in Chile"*

> *"You even have risks inside your house, well,…the thing is, one has to take precautions."*

An essential element of risk perception is the attribution of the causes and culprits. This is nothing new in scientific discourse, and Douglas and Wildavsky (1983) have already dealt with this element. Even within the risk research community, it is common to use economics and politics as simple explanations for complex disasters in a heuristic manner (Oliver-Smith, 2013).

3.5.2.1.3 Psychological Processing and Emotional Strategies

Cognitive dissonance. When examining the mechanisms of how risks are psychologically processed as potential threats and catastrophic events that have occurred, the strategy of ignoring to resolve cognitive dissonance emerged.

"If we were worried about it, we would not be able to work peacefully.."

Feelings. Retreats into religion, asking for salvation, and the hope that nothing bad will happen were observed. We also observed black humor, which we interpreted as a way of dealing with psychological stress.

"The guy had to run faster, but he was: (laughing) uh uh... he was cuckoo+ yes, he saw how many people were dying, and he was precisely in the Aysén Fjord"

Security paradigm. The overall demand for safety also influences attitudes towards individual natural hazards.

"the Chileans react saying, "ah, yes, it's shaking", without much concern"

Findings in the categories of emotions and psychological processing primarily pertain to psychological research. Psychological models typically explain the emotional aspects of risk perception. However, their exact applications would lead us away from focusing on this study. For example, the statement "if we constantly think about it (natural hazards, author's note) we would not be able to work in peace" can be related to cognitive dissonance theory and describes the strategy of ignoring (Festinger 2001).

The behavior of the social environment has a decisive influence on which risk decisions are made, especially under the condition that a *pareto optimum* must constantly be found between different risks (Dannenberg et al., 2022).

3.5.2.1.4 Territorial and Professional Identity

Incorporation of hazards. Dealing with dangerous situations has become a part of the identity fishermen in the area, whether traditional or industrial, because they must use suitable technical measures and personal skills to constantly deal with the forces of nature during their work. This is also strongly tied to the landscapes in which they work and identify.

3.5 Results and Discussion

"But that is the danger we have. (fast) But that is the danger that we fishermen have::+ assumed. It was what we liked, e:h,.. (fast) and we keep at it and make money, therefore we are always going to be there in between it all, without considering the consequences"

"But it is not our fault, (&) we were already here"

Responsibility for preparedness. We observed that preparation for dangerous situations and disasters was seen as a personal responsibility.

"You even have risks inside your house, well,...the thing is, one has to take precautions."

"...as fisherman ourselfs we are prepared for-...for this. when this, as I told you, this-:.. these natural forces"

Gender roles. Traditional gender roles, such as men undertaking risky tasks, have been observed to influence the incorporation of risks.

"...but he has enough to feed his family comfortably"

We propose *territorial and professional identity* as concepts that relate to aspects such as the influence of habitus or the incorporation of the concern of the functional system in which people are included. Thus, they correspond to approaches to risk perception in sociology, social psychology, and other related disciplines. The corresponding risk framework would be, for example, a serial amplification of the risk (Kasperson et al., 1988). Empirical work in these areas deals with, for example, place attachment (Dominicis et al., 2015; Masuda and Garvin, 2006), discourse (Olofsson et al., 2016), habitus(Hernández Aguilar and Ruiz Rivera, 2016), or inclusion in a social subsystem (Luhmann, 1991).

3.5.2.1.5 Perceptions of the Nation-state

Disaster Management and Understanding of the Nation State. We rarely observed references to formal disaster management in questions about how to deal with such events. There is deep mistrust of the state, including disaster response organizations such as the National Office of Emergency of the Interior Ministry (ONEMI). Overall, the respondents hardly identify with the Chilean nation-state, and the lack of representation goes beyond current politics.

> "True, (&) we know how they manage, but for the same reason, if there is a volcano that is going to explode, (&) and the government knows that it is going to explode at such and such an hour, they are not going to tell the community, & right, only, yellow alarm, nothing else".
>
> "I think that if I was a politician they would have killed me by now"

Search-and-Rescue (SAR) and Health Infrastructure. There is high awareness that SAR and health infrastructure are out of reach.

> "And, ... and also being so far away sometimes causes you...ehm, ... insecurity even:: at a personal level, for your health. Because, for example, if today, today the weather is bad, today no boat has entered, and if someone is sick with appendicitis, apendicitis or has:, has any disease, no boat is going to take him out"

Pioneers, Frontiers and Centralism. The territorial identity mentioned above is also closely linked to the pioneering ideas among the descendants of *white* settlers.

> "We are building our homeland out there, and the other important thing, we are building sovereignty, which is important. And who is helping us? Who? (break) There it is, because of the centralism."

Global Economy and Financialization. Global economics plays a relevant role for fishermen, as they can barely compete with transnational economic players, and options to sell in local markets diminish.

> "...salmon farming is around here because of the bucks"
>
> "I don't know, sometimes one-.... as I tell you, & I don't know what to do, & I don't know whether to place myself on this side, & I'm already on this side, I don't know whether to change, to fight, I don't know, ... I don't know. For now, I need to stay, because I need the money"

Mistrust and Conspiracy theories. One perspective observed is that state institutions are not trusted, and politicians are accused of deliberately withholding information about hazards.

> "If there is a volcano that is going to explode, & and the government knows that it is going to explode at such an hour, they are not going to tell the community, & right, and only announce a yellow alarm, nothing else."

3.5 Results and Discussion

"I worked in cultivation for years, so I know:: how the law is violated, I know... that the authorities turn a blind eye or look the other way, & knowing it is being done. I know it happens, and (angrily) it is annoying. (...) But in an interview like this one, I can say it."

"...until political interests got involved again, and now the masonic (laughing) plan is underway again."

Institutions, organizations, and procedures for disaster mitigation and management are usually nation-state based. Therefore, we consider it essential to understand the perspectives of those affected by natural hazards towards the state and its institutions, especially when it comes to a study area with colonial history.

The fishermen in the study area showed no trust in risk management institutions and even felt completely disconnected not only from politics but also from the national state itself. The phenomena in our case study are not singular; subsistence farmers in Chile also feel unrepresented by the state, regardless of whether they have a Criollo or indigenous identity (Braun, 2021a; Di Giminiani, 2016a). In La Paz, Bolivia, Nathan (2008) also found that there is basically no link between affected people and formal risk management; it is a social practice and a way of life to solve one's problems alone or, at best, within the family.

It is also common for top-down environmental planning, including disaster mitigation strategies, to generate resistance at the local level and fuel conspiracy theories spread by actors with economic interests (Hurley and Walker, 2004). Evidently, some of the participants' statements cannot be taken literally, especially when assumptions about how distant powerful actors are described involve xenophobic, paranoid notions, and support conspiracy theories. However, the specific institutional processes and circumstances that make the lives of subsistence fishers difficult and marginalize them can certainly be judged by no one better than themselves. It is the role of science to translate the testimony of the people concerned, not to judge whether the attributions of the causes of these processes are correct or wrong(Czarniawska-Joerges, 2009, pp. 51–52). In our empirical case, fishermen perceived their resilience as threatened by the environmental impacts of aquaculture.

The reconstruction phase after a natural disaster is used to impose the interests of powerful economic elites, sometimes even though there is a participatory process. After the 1985 earthquake in Chile, nationally controlled disaster recovery in Santiago de Chile was used to implement neoliberal infrastructure (González, 2022). In New Zealand, there were similar reports on reconstruction after the 2012 earthquake. The citizens felt left out even though citizen participation was present in this case (Cretney, 2019). In the regulation of environmental issues at global level, too, the strategic interests of powerful players often prevail (Cairns and Stirling, 2014).

The fact that (both spatially and socially) distant actors make deliberate decisions that affect the living conditions of small economic actors is a truism that is subject to the social and political sciences and is generically described in the context of DRR, for instance, in the PAR model (Boillat et al., 2020; Wisner et al., 2010). Rather than dismissing statements as conspiracy theories, it is better to recognize that it is a hallmark of postcolonialism to view the nation-state as the only legitimate form of structure and organization (Arora and Stirling, 2023; Cairns and Stirling, 2014; Quijano, 2000).

We advocate a more abstract approach to attitudes towards the state in the context of risk perception studies and focus on social vulnerability perception. We must translate these statements into indicators of how the root causes and pressure factors generate vulnerability for a specific group. It will have to be accepted that attributions are not initially plausible from a scientific perspective, but we should nevertheless recognize the concerns behind them.

3.5.2.2 Enabling More-than-human Visions by Renouncing the Insistence on the Distinction between Natural Hazards and Environmental Degradation

Many of those affected are concerned with how nature is appropriated, the accumulation of profits from its use, and the distribution of profits. However, this is not observed only at the superficial level in exchangeable goods. The displacement of traditional sustainable practices is also linked to this, as well as the need for local communities to have a greater say in local common-pool resources (or those that existed before colonization).

3.5.2.2.1 Environmental Transformation

"The The industry impacts more than nature"

The statement demonstrates, that natural hazards and anthropogenic impacts on ecosystems have not yet been distinguished. The environmental degradation described included resource extraction, contamination (chemical and optical), deforestation, and invasive species.

"Now, from my perspective, I think that the earthquakes, tremors, and so many other issues are effects… of oil extraction"

"For example, in the old days here in the region, there was a lot of octopus and small fish. The salmon wiped that out.."

"why did we put the aquaculture industry (pounding on the table) on top of the natural shoals?"

"Because this is a matter of all mankind. Look. Don't forget that '86 was the first convention on the issue of ecosystem balance, not right, in.., uh. Kyoto." (...)

"this place was located in a point of equilibrium between the tectonic plates, which implied we were never going to suffer.:, e:h, earthquakes, because there was an equilibrium of the plates. And it happens that-... we were saying, but what happened to the equilibrium, and a professor from the University of Chile, ... said, (quickly) the equilibrium was broken."

These statements are just a few examples that illustrate that there is no differentiation between natural hazards and environmental degradation, as is done academically, and above all in the natural sciences. Therefore, this distinction was an incorrect reference frame. An investigation into risk perception is inadequate if the perception of environmental change is excluded.

3.5.2.2.2 Human-Environment-Relations

Nature is assumed to react to illegitimate human use, although this may not occur directly. These reactions may be events described in scientific knowledge (particularly in the natural sciences). There is not necessarily a distinction in the perceptions of fishermen. In particular, the hybrid concepts show that the interviewees construct few or no strict cause-effect chains in the way Western science would describe the world. It is noticeable that no distinction is made between natural hazards for humans and the consequences of the anthropogenic use of natural resources for nature, particularly the degradation of ecosystems. Rather, 'nature' is seen as an entity that reacts to perturbations like an organism, and this can also be indirect. Nature responds to the way humans treat it, which can be disjunctive in terms of time, space, or phenomena.

The implicit assumption is that natural processes determine, and should determine, human behavior, including the manner and intensity of natural resource use.

Participants' attitudes towards natural hazards are closely linked to their perception of the landscape and natural space, which is not seen as a hazardous area but as a source of livelihood, landscape aesthetics, and part of their identity. They also refer to nature, giving signals about what is about to come and what animals do, and humans should listen to them.

> *"Nature is perfect, ..."*
>
> *"Practically 20 days before, we noticed that there were no conger eel, the conger eel had already escaped. Look, you see that—on the subject of fires, it is true, animals have a sixth sense, they perceive, ... the stubborn one here on earth is man"*
>
> *"What we have to do now is learn how to coexist with nature better and do as little damage as possible"*
>
> *"No one commands you, you have no color, no religion, no politics, ... timetable. Me, the day starts when the sun rises... and when—that's a day, it's twelve o'clock, what does that mean, I don't know (quickly) The day starts when I want it to. + And it ends when I don't want to work anymore. That's it. Freedom."*

3.5.2.2.3 Appropriation of Nature and Ownership structure

Attachment to the territory goes hand-in-hand with the claim of the legitimate use of natural resources by traditional fishermen and fjord dwellers, while industrial production with foreign investment is rejected. In salmon farming, industrial resource extraction is considered illegitimate because it interferes heavily with natural processes.

> *"There are no long-term considerations in the industry. They exploit today without thinking about what I'm going to do tomorrow"*
>
> *"why did we put the aquaculture industry (pounding the table) on top of the natural shoals.."*
>
> *"How come each one of the animals makes its territory and- and, and respects other territories, but we don't."*
>
> *"And since we are the owners of our own work, eh, but, l::- the other party that has the job is the one that has to...has to see how it is going to do it in case there is a hecatomb...."*

Formal ownership hardly plays a role in emergencies. Boats that belong to companies or authorities, for example, were made available informally and free of charge for search-and-rescue and food supply purposes in 2007.

3.5.2.2.4 Displacement of Sustainable Traditions

Some interviewees emphasized traditions and local social cohesion, especially within families. One aspect is that traditional knowledge is passed on from generation to generation.

> "the other one doesn't have the money, but he has enough to eat for his family."

The displacement of traditional livelihoods is perceived as a risk.

> "El gobierno, lo que era Sernapesca, prácticamente no existía, existía a nivel de pescadores artesanales, pero no para la industria del salmón."

3.5.2.2.5 Regional Common-Pool Resources

One phenomenon observed in several interviews is the relevance of regional solidarity, which states that spatially absent actors influence their local ecosystems. It is unacceptable that local and regional resources are not managed by local people and that benefits and profits go to external profiteers.

> "So we acknowledge this weakness, and we think that here, uh, uh, if you have accurate mapping of the Liquiñe-Ofqui Fault and its fragmentations, we should identify whether there is a fault below. Detecting the absence of such fault lines here could be crucial for future events like, I don't know, earthquakes or something like that, and if you have suitable terrain... to... accommodate this, it could enhance our utilization of resources, both for tourism and as a productive system."

In general, the fact that there are few conversations or agreements regarding decisions plays a particularly important role. Direct, personal, and verbal communication, in particular, has very high value.

> "It seems like there is no... there is no will of the industry to: .... (quickly) in getting along with the neighbors, .. then, maybe that is the worst mistake, not wanting to get along with the neighbors. (...) The neighbors are the locals or the other economic activities.."

3.5.3 Hybrid Understanding for Symmetrical Relations

The concept of emergent order reflects the perception of risks against the inherent ontology behind the various scientific disciplines that have constructed approaches and theories of risk perception. However, some statements and attitudes were not considered in this systematic review because they would not even be interpreted as perceptions of natural hazards due to a different ontology and were excluded from the data analysis. This eliminates the distinction between natural hazards and environmental damage since this distinction is not made by laypersons but is based on the epistemology of academic knowledge production. This goes hands-in-hand with the hegemony of Eurocentric knowledge production. If we allow for ambiguities to overcome the nature/culture divide, a hybrid understanding can emerge that enables the analysis and understanding of risk perception. In addition, the findings can be connected to other research fields, disciplines, and knowledge forms, and vice versa.

Generally, perceptions of natural hazards are interwoven with perceptions of the environment. Risk perception research should respect the fundamental difference between the Western ontology of nature and Mapuche Cosmovision, in that the landscape is not divisible in assets (Carruthers and Rodriguez, 2009b; Le Bonniec; Neira Ceballos et al., 2012; Ranjan et al., 2021a). According to natural science ontologies, the division of nature into compartments and assets also makes it impossible to distinguish between natural hazards and environmental destruction (Sullivan, 2017). Therefore, we consider it important to focus on environmental discourse in risk perception research. In the study area, environmental discourse has moved into an area of tension between development and conservation (Bachmann-Vargas and van Koppen, 2020); similarly, corresponding phenomena have been reported in Botswana by Dintwa et al. (2022). The risk perception of natural hazards is closely interwoven with the ontology of nature. By acknowledging the entanglement of environmental discourses, we can develop a more nuanced understanding of risk perceptions. By allowing ambiguity and hybrid meanings, it should be possible to connect to indigenous knowledge systems and grasp the actual difficulties people face in their daily lives.

A closer look at and review of other articles on risk perception reveals similar patterns in other empirical cases. An example from Chile is the following: in the Huasco Valley in northern Chile, the community is concerned that the heavy metal contamination of rivers is caused by leakage in mining tailings due to landslides. Geochemical analyses show that the dissolved heavy metal ions correspond to the geogenic ("natural") background load and that their concentration was not increased by the landslides. However, the concentration of dissolved heavy metals reacts to climate anomalies with high precipitation, which are most likely to occur because of anthropogenically induced climate change (Valdés Durán et al., 2022).

In this respect, the suspicions of the population that human activities influence the situation are justified, albeit via other cause-effect pathways.

In a corresponding way, hybrid understanding concepts (environmental degradation, human-environment relations, appropriation of nature, displacement of sustainable traditions, and regional common-pool resources) applicable to other countries, such as research in Peru (Klimeš et al., 2020), Brazil (Bustillos Ardaya et al., 2017), Ghana (Adomah Bempah and Olav Øyhus, 2017), and Australia (Ali et al., 2021).

We deliberately decided against limiting the concepts to Indigenous perspectives and specified them in this manner for several reasons. First, the non-indigenous interviewees did not necessarily show a nature-culture distinction in the sense of differentiating between natural hazards and environmental degradation, as outlined earlier. Second, and more importantly, framing the hybrid concepts as indigenous perspectives would be the second external distinction of modernity, as described by Latour (Latour, 1995), the distinction between modern and "pre-modern" cultures. Thus, we differentiated between different cultures and reproduced the asymmetrical relationships.

3.6 Conclusions and Outlook

In this study, we explored how the risk perception of natural hazards can be investigated and described in a decolonized manner. Adopting an actor-network theory (ANT) approach was essential because it has previously facilitated the decolonization of research on disaster risk reduction, allowing the inclusion of more-than-human perspectives. This meaning that this approach can also enable a symmetrical exploration of perception in a corresponding manner (Gaillard, 2019a; Neisser, 2014).

In narrative interviews in the Chilean Patagonian Fjordlands, we found many aspects of risk perception that corresponded with insights into risk perception from various scientific disciplines, thus representing the emergent order of academic knowledge production. However, many statements do not fit this order at the ontological level. Against the background of actor-network theory, we realized that the nature-culture divide manifests itself in this field of research in the separation of natural hazards and environmental degradation. Laypersons do not necessarily make this distinction. At first glance, one might think that these are pre-modern perspectives. However, this is the precise pitfall that leads to asymmetrical relationships, as described by Latour (1995). This would mean a distinction between different cultures, consequently reproducing coloniality through the hegemony of Eurocentric knowledge production and providing the mental foundation for accompanying epistemic violence.

This brings us to the conclusion that research on the risk perception of natural hazards should be developed starting from the perception of the environment and human-environment relations and not from the perception of (general) hazards (see also Chap. 2). We suggest thinking and describing risk perception less under the precondition of what is scientifically identified as a "natural" hazard. In addition, to widen this perspective, empirical studies should place more emphasis on social and socio-environmental vulnerability perceptions (Abid et al., 2016; Antronico et al., 2020; Dintwa et al., 2022; Morelli et al., 2022; Tanner and Árvai, 2018).

In view of the fact that six out of nine ecological planetary boundaries have yet to be exceeded, and most natural hazards indeed show anthropogenic influence (Gill and Malamud, 2017; Heinberg, Richard and Miller, Asher, 2023; Richardson et al., 2023; Schröder, 2023), we ask ourselves to what extent the ontological separation into natural hazards and environmental damage and the resulting separation into different fields of research on their perception still makes sense. In a stochastic manner, it might be possible to attribute the anthropic share to disasters (Kreienkamp et al., 2021). However, what relevance does this have to the actors involved and people affected in their daily lives and livelihoods? Instead of discussing whether physical hazards occur with or without our influence, it is perhaps time to put more effort into developing methods and procedures that consider how collective action can be designed in such a way that ensures resilience among the regional assemblages of human and non-human actors.

People who live directly from and with the natural environment are perhaps less concerned about natural hazards than about human-environment interactions that endanger landscape resilience. They need the co-production of knowledge about the landscape and the co-creation of land use (Sieber and Cybele, 2024). Ultimately, integrating these perspectives can contribute to more resilient landscapes.

The full consequences of epistemic emancipation through thinking outside the box of academic knowledge production (which corresponds to the existence mode Reference [REF] in ANT terminology) and accepting pluriverse knowledge systems in natural hazard risk perception research are yet to be developed. The next task for science is to establish more connections between academic knowledge and local ecological knowledge (El-Hani et al., 2022; Renck et al., 2023) and develop governance systems that consider these corresponding (indigenous)

3.6 Conclusions and Outlook

visions (Bavikatte and Bennett, 2015; Chen and Gilmore, 2015; Ranjan et al., 2021a). Further studies should analyze which synergies, ambiguities, and incommensurabilities arise when the proposed hybrid concepts of risk perception in more-than-human assemblages are applied to risk and land use governance, and how they can be connected to new legal approaches, such as the rights of nature or biocultural rights (Bavikatte and Bennett, 2015; Gutmann, 2023).

4 Navigating a Trojan Horse in the Last of the Wild: Pine Trees, Agroforestry, and Land Zoning Assemble the Landscape Resilience Dilemma in Patagonia

This chapter is a preprint of an article that is currently in the review process.

Abstract

The resilience of landscapes is increasingly compromised, even in regions with minimal human influence, such as cool-temperate forests. Deforestation, climate change, and natural hazards are threatening these last wild areas, endangering biodiversity, carbon-storing soils, and rural livelihoods. Empirical case studies provide critical insights as an "early warning system," to help anticipate land use risks and identify strategies to enhance landscape resilience in the face of these challenges.

For example, in the Chilean Patagonian region of Aysén, forest plantations with exotic tree species and spatial planning with land zoning are intended to ensure landscape resilience. With the help of planning evaluation, using a combination of conformance and performance approaches, we analyze whether this set of instruments can address the risks of biodiversity and soil loss, land-use-related natural hazards and disregard for (indigenous) risk perceptions.

Analyzing the land zoning critically, we argue that eventhough regional spatial planning in Aysén is dedicated to sustainable land use, land zoning is not in conformity with its environmental objectives of eradicating invasive species, is expected to perform low in mitigating erosion and natural hazards, and is not addressing local risk perception. This is particularly evident in the fact that the resilience potentials that agroforestry practices could offer cannot unfold. The agroforestry concept is used as a land use category in land zoning to promote the expansion of further forest plantations with exotic tree species, conflicting with local risk perception and traditional rural livelihoods. Thus, agroforestry becomes a trojan horse, with limited or even counterproductive effects on landscape resilience.

Inspired by the polycentric governance paradigm, we propose three alternative strategies: promotion of traditional agroforestry under exclusion of exotic tree species, land zoning categories for disaster risk mitigation co-created by local communities, and promoting community forest management.

4.1 Introduction

The global community has recognized the urgency of a sustainable transformation and global environmental goals are being agreed on (Cernev, 2022; Fukuda-Parr and Muchhala, 2020; Galaz et al., 2012). In order to achieve these goals, it is necessary, to scale them down to landscapes, which allows for the concretization of sustainability goals and measures, as well as the identification of conflicting goals and trade-offs (McDermott et al., 2023; Nilsson and Persson, 2012; Pérez Piñán and Vibert, 2019). Thus, sustainable landscapes are a key lever for sustainability (Chan et al., 2020; Jeetze et al., 2023; Morrison et al., 2022). The sustainability of land use and the interactions of different land uses at the landscape scale have recently been discussed under the concept of landscape resilience (Beller et al., 2015; Cumming, 2011b; Plieninger and Bieling, 2012; Rastandeh and Jarchow, 2022a; Schmidt, 2022). While the concept is still contested in detail, it generally refers to the capacity and the ability of a landscape to withstand and recover from disturbance while maintaining its essential structure and functions, including necessary feedbacks (Rastandeh and Jarchow, 2022a; Schmidt, 2022, p. 6). It is argued that the ecological resilience of landscapes is closely linked to socio-ecological and socio-economic resilience (Beller et al., 2015). Thus, the importance of certain forms of land use for maintaining socially desirable environmental conditions, such as agroforestry, comes into focus (Lomba et al., 2020; Plieninger et al., 2023; Rudel and Meyfroidt, 2014). In this respect the concept of landscape resilience considers society and landscape not as individual systems, but as intertwined human-environment networks, making it also connectable to conceptualizations of more-than-human worlds (Head, 2012).

4.1 Introduction

Biggs et al. (2012) have proposed principles for the ecosystem resilience, which Rastandeh and Jarchow (2022b) have discussed for appropriate application and translation to landscape resilience governance (Biggs, 2015). Spatial planning can contribute to maintaining diversity and redundancy, and managing ecosystem connectivity with tools such as land zoning (Metternicht, 2018). Spatial planning can also make valuable contributions to the resilience of landscapes against natural hazards (Greiving, 2002). Depending on the planning system, spatial planning influences the type and quality of the respective land use, such as regulating the density of settlements (Datola, 2023). In the case of rural land use, however, the factors that regulate land use are usually the responsibility of sectoral planning bodies such as government agencies for forestry and agriculture (França et al., 2022).

Another principle of landscape resilience proposed by Biggs et al. (2012) is polycentric governance (PG). Factors have been identified, that make the sustainability of natural resource use more likely (Ostrom, 1990). These design principles have been reconstructed from hundreds of empirical cases (Cox et al., 2010). Clear user and resource boundaries, proportional equivalence, collective choice arrangements, monitoring, graduated sanctions, conflict resolution mechanisms and local autonomy significantly increase the likelihood of sustainable natural resource use (Baldwin et al., 2023; Cox et al., 2010; Goetz et al., 2023).

Forests play a particularly important role in landscape resilience (Hetemäki and Seppälä, 2022; Popp et al., 2012; Stubenrauch, 2022). Although they often receive little attention in public discourse, continental and boreal forests in particular are highly relevant to the global climate as carbon sinks (Gauthier et al., 2015; Martínez Pastur et al., 2022; van Assche et al., 2022; Watson et al., 2016). Many of these forests have been largely unaffected by land use, making them some of the last of the wild areas (Sanderson et al., 2002). However, deforestation and land use change are increasing (Moreira-Dantas and Söder, 2022). A typical land use sequence thereby is:

- deforestation for agricultural use, often with the displacement and exclusion of local users from the appropriation of the forest resources (Ramcilovic-Suominen et al., 2022);
- intensive land use with monocultures, leading to biotic homogenization (agricultural or forest plantations) (García et al., 2018);
- last but not least, partly reforestation as semi-natural forests or tree plantations. However, afforestation comes with its own risks for landscape resilience, regardless of the fact that it is commonly labeled as sustainable by institutional investors and in the media discourse (Braun, 2022; North et al., 2019).

Fig. 4.1 Graphical abstract of the case study. (The aim of the study is to investigate whether spatial planning can mitigate risks to landscape resilience, for instance by land zoning, and to develop alternative governance strategies. The study area Chilean Patagonia as an example of a last of the wild area. The theoretical framework and methodological approach is planning evaluation, while the development and discussion of alternative strategies is inspired by the design principles from normative polycentric governance)

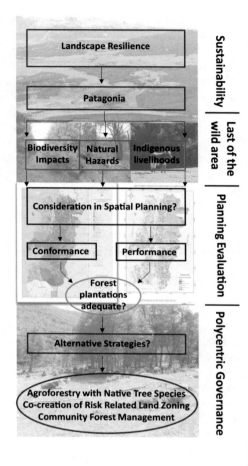

Landscapes of these temperate "last of the wild" forests are threatened by this land-use succession, and maintaining—or restoring—their resilience should be a high priority (Araujo and Austin, 2020; IPCC, 2023; Thurner et al., 2014). However, there are many question marks as to how this could be achieved (Aguiar et al., 2021; Dlamini and Montouroy, 2017; Moreira-Dantas and Söder,

2022; Newig et al., 2019; Smith and Stirling, 2008). One possible governance approach used in different regions is spatial planning and the land zoning tool, which is the focus of the study (Ruiz Agudelo et al., 2020; Saunders and Becker, 2015). Based on the results of the planning evaluation, we develop proposals for land use governance, including but not limited to spatial planning instruments (Hull et al., 2011). In doing so, we draw on normative polycentric governance to address leverage points for resilient landscapes (Cox et al., 2010; Davoudi, 2003; Goetz et al., 2023; Ibrahim et al., 2023; Martin et al., 2022). These issues are assessed and discussed using the example of the Chilean Aysén region (Allan et al., 2017; Inostroza et al., 2016; Sanderson et al., 2002).

4.2 Study Site

The study area is the Aysén region of Chile (Fig. 4.2) (45°34'16.3"S; 72°04'06.7"W). It is the second most southern region of Chile and is located in the Patagonian part of the South American continent.

In the Aysén region, intensive land use began just over a hundred years ago (Fajardo et al., 2022). Various indigenous communities have lived here for thousands of years, modifiying the landscape through the use of fire for hunting, but a higher population density did not occur until the immigration of the Mapuche (indigenous communitites of southern Chile) at the end of the 18th century. They were forced to flee from south-central Chile from the genocide of the "pacification" campaign and the colonization of their ancestral lands (Belardi, 2021; Garvey et al., 2023). From the 1920s to the 1960s, land use in Chilean Patagonia was then promoted by the Chilean state through the granting of land titles for agricultural cultivation, which marked the beginning of deforestation and land use (Rasmussen and Figueroa, 2022). The deforestation necessary for this was mainly carried out by slash-and-burn (Fajardo and Gundale, 2015). Formal land ownership has been predominantly acquired by "white" farmers who migrated from central Chile or Europe to seek their fortune here. The Mapuche who have fled to this region were once again institutionally discriminated against and marginalized (Pérez and Walter, 2020; Rasmussen and Figueroa, 2022).

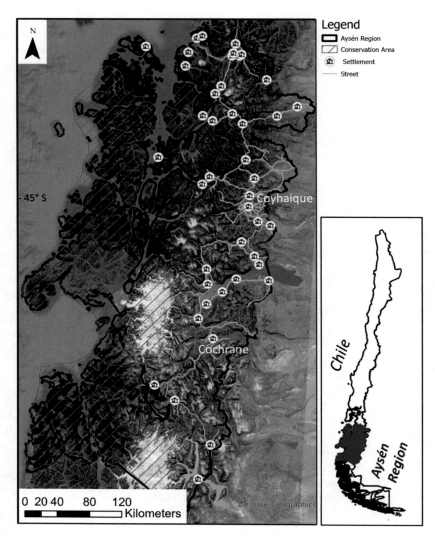

Fig. 4.2 Map of the study area. (Administrative boundary of the Aysén region in black, with shaded relief and land cover in true color satellite imagery. National protected areas cover large areas, especially in the west (purple stripes). The Andean mountain range is covered by dense primary forests. Deforestation and extensive land use are mainly to be found along the road network and especially in the transition zone from Patagonian forests to the steppe in the east, for example around Coyhaique (~45° S) and Cochrane (Cartography: Danny Tröger; Data: ESRI, NASA, USGS ASTER GDEM, Gobierno de Chile))

4.2 Study Site

The intensity of subsequent extensive land use (sheep and cattle ranching) led to widespread ecological problems on the highly erodible volcanic ash soils, such as loss of fertility and desertification, changes in the sediment load of rivers, displacement of floodplains and frequent landslides (Hernández-Moreno et al., 2023). In the 1970 s, reforestation efforts began in the national park near the regional capital of Coyhaique with coniferous tree species alien to South America, such as *Pinus contorta* and *Pinus ponderosa* (Braun et al., 2017b; Fajardo and Gundale, 2015). This was part of the central government's decision to subsidize such afforestation with fast-growing species through the decree 701 (Heilmayr et al., 2020). This erosion control incentive was a spatially effective, but not spatially coordinated, land-use management tool. Entire landscapes in south-central Chile, but also in some areas of Patagonia have changed as a result of this policy (Bopp et al., 2020; Pérez and Simonetti, 2022). The establishment of tree plantations has resulted in three socio-ecological risks to landscape resilience, which are the focus of this article:

- First, the proliferation of exotic tree plantations with invasive species that affect plant and animal biodiversity of plants and animals and do little to reduce soil erosion (Braun et al., 2017b; Bravo-Monasterio et al., 2016; Fajardo et al., 2022; Heilmayr et al., 2020; Huertas Herrera et al., 2022; Paritsis and Aizen, 2008),
- Second, inadequate mitigation of natural hazards such as landslides and the increasing risk of forest fires, especially under the expected effects of climate change (Braun et al., 2021; Olivares-Contreras et al., 2019; Sandoval et al., 2014; Soto et al., 2023; Tröger et al., 2022),
- Third, the violation of the territorial identity of the indigenous peoples in particular and the displacement of traditional land uses. This leads to the abandonment of many areas, which favors green land grabbing (e.g. outdoor tourism or forest plantations), which in turn can lead to environmental conflicts (Braun, 2021b; Carranza et al., 2020; Carruthers and Rodriguez, 2009a; Funk, 2012; Latorre and Pedemonte, 2016; Louder and Bosak, 2019; Mendoza, 2023; Mendoza et al., 2017; Ranjan et al., 2021b; Salazar and Riquelme Maulén, 2022; Tröger and Braun).

Some spatially effective governance incentives and instruments for land use, such as subsidies for forest plantations under Decree 701, but also the network of national nature reserves, date back to the Pinochet dictatorship and were highly centralized, i.e. hierarchically organized (Arriagada et al., 2018; Bopp et al., 2020; Carruthers, 2001). In the course of Chile's decentralization, spatial planning instruments have been introduced in the last two decades at the subnational level over the last two decades (Vicencio et al., 2023). These include the creation of a regional plan of territorial order, land zoning and a spatial plan for the prevention of natural hazards (Ossandón et al., 2020; Vicencio et al., 2023). The extent to which these instruments can address the three risks to landscape resilience will be assessed using a mix of planning evaluation methods (Alexander, 2011).

4.3 Materials and Methods

The theoretical foundations and methodology of planning evaluation are used to analyze the practice of spatial planning and land zoning (Alexander, 2016; Alexander, 2006; Allmendinger and Tewdwr-Jones, 1997; Hull et al., 2011). Planning evaluation distinguishes between conformance and performance approaches (Alexander, 2011; Dadashpoor and Heydari, 2022; Feitelson et al., 2017).

The conformance approach assess the extent a plan is coherent and the corresponding outputs and outcomes are (or can be, if ex-ante) consistent with the plan. The purpose of the conformance approach is to answer the following main questions (Alexander, 2011, p. 41): "Do the outcomes 'on the ground' match the descriptions?" and "Do the tools deployed to implement a policy or plan, in fact work to promote its espoused objectives?" From these general guiding questions, we develop the specific questions listed in Table 4.1.

The documents taken into account for the conformance evaluation are the Regional Plan "**P**lan **R**egional de **O**rdenamiento **T**erritorial" (hereinafter "Regional Plan" or **PROT**), published in 2013, and the corresponding Plan for the Prevention and Protection against Natural and Anthropic disasters "Plan de Prevención y Protección ante Situaciones de Emergencia y Desastres Naturales y Antrópicos" (hereinafter "Disaster Prevention Plan"), latest version published in 2019.

Performance analysis in planning evaluation refers to the extent to which the plan has an impact on the decisions and implementations addressed by the plan, such as the subordered regulatory plans, and correspondence with other spatial instruments, such as national parks, or with sectoral policies, such as forest policy (Alexander, 2011; Mastop and Faludi, 1997). In order to determine this, grey literature (government documents) was examined to determine whether they referred to the PROT or to the natural hazard prevention plan.

4.3 Materials and Methods

Table 4.1 Overview of the operationalization of planning evaluation. (The evaluated documents are the Regional Territorial Plan, with the Land Use Zoning Plan and the accompanying document Disaster Prevention Plan)

Evaluation approach	Aspect	Guiding questions for coding
Conformance	Conformity of land zoning plan to prescriptions	Does the land use zoning conform to the environmental goals and objectives set out in the regional plan? Does the zoning support identity goals (land use aspects)?
	Correct instruments to achieve desired outcome	Can the land use zoning achieve the environmental goals and objectives? Can the plan support identity objectives with spatial relevance?
	Conformity between disaster prevention plan and land use zoning	Does the land use correspond to the information in the disaster prevention plan?
	Conformance of disaster prevention plan with risk perception of vulnerable groups	Does land zoning support indigenous people's risk perception?
Performance	Influence on subordered Plans and relevant policy instruments	Is the PROT mentioned in relevant documents or known and referred to by experts/staff of the relevant institutions?
	Correspondence with other (spatial) instruments	Are other spatial instruments consistent with the land use planning? Is the PROT mentioned in policy documents or by experts in relevant positions?

For the operationalization of the evaluation questions, the land use zoning of the Regional Plan of Aysén (PROT) is analyzed and evaluated for conformance with its environmental and identity objectives. The focus of the analysis of the Disaster Prevention Plan is on the land-use planning relevant risks landslides and wildfires (Felgentreff, 2003; Greiving, 2002). Both documents were subjected to a content analysis, with deductive coding to extract relevant sections (see Appendix). Coding was derived from the questions in Table 4.1 and included environmental goals and objectives, identity and indigenous community issues, land use and zoning categories, soil, vegetation and land-use change related codes.

The assessment is complemented by a spatial analysis with geospatial methods (ArcGIS Pro 3.0.3). Land Zones for preffered agroforestry use from the land zoning map were superimposed on base maps (boundaries, streets, land cover data) provided by the regional government. *Pinus* Plantations were derived from the results of Hernández-Moreno et al. (2023).

4.4 Results

First, a comprehensive presentation of the goals and objectives of the regional territorial plan ("PROT") is given in Sect. 4.4.1, followed by the land-use related contents of the corresponding natural hazard prevention plan in Sect. 4.4.2. In Sect. 4.4.3 presents the land use zoning, including the geospatial analysis. This information then leads to the conformane evaluation in Sect. 4.4.4. The results section concludes with the performance evaluation in Sect. 4.4.5.

4.4.1 Goals, Objectives and Action Plans of the Regional Territorial Plan

There are eight general strategic goals in the Regional Plan, called macro goals, of which #5 Environment (next section) and #8 Identity (Section 4.4.1.2) are directly relevant to landscape resilience and the focus of this study.

4.4.1.1 Environmental Goals and Objectives

Macro Objective #5 of the Regional Plan emphasizes the importance of environmental conservation and sustainable use, and specifically highlighting the reduction of erosion in rural areas (División de Planificación y Desarrollo Regional, 2013, 5–7).

The action plan part distinguishes between actions for native forest management and plantation forest management (División de Planificación y Desarrollo Regional, 2013, p. 363). The action plan for native forest management focuses on the promotion of bonuses, the dissemination of best practices and guidelines, and certification.

The **forest plantation paragraph** includes two aspects: the first one is to promote **afforestation in preferential zones for reforestation (a category that does not exist in the land zoning, see Sect. 4.4.3)**, and the second is to promote **forest plantations with agroforest** use, with financial support to the owners financially for the period in which livestock has to be kept out of the tree growing areas until trees are large enough (División de Planificación y Desarrollo Regional, 2013, p. 363).

The action plans propose as strategies to achieve the following environment goals: **eradication of invasive species** and **reduction of erosion in rural areas** (División de Planificación y Desarrollo Regional, 2013, p. 369):

Other land use objectives, such as the location of landfills, are relevant to regional planning and landscape resilience but do not affect the large-scale land uses that are the focus of this study.

4.4.1.2 Identity Goals and Objectives

Macro Objective #8, Cultural Heritage and Regional Identity, deals with the preservation of cultural heritage, the recovery of **territorial identity, inclusion**, the rescue and highlighting of **indigenous peoples' culture**, and equal opportunities (División de Planificación y Desarrollo Regional, 2013, 8–9).

The action plan for the identity objectives proposes a "diagnosis of the current situation of native peoples in Aysén". "The processes of strengthening identity will take into account the objectives of adapting **modernization processes** to development and strengthening equality of opportunity" (División de Planificación y Desarrollo Regional, 2013, p. 370).

4.4.2 Disaster Prevention Plan

The Natural Disaster Prevention Plan classifies forest plantations with exotic species as high risk areas for wildfire (Gobierno Regional de Aysén, 2019, p. 103). An attached map highlights the plantations as wildfire areas (Gobierno Regional de Aysén, 2019, p. 115).

The document also highlights the role of vegetation in landslide mitigation, and there is a map of landslide risk zones that includes zones without vegetation as a risk factor (Gobierno Regional de Aysén, 2019, p. 76). Forest plantations and native forests are not distinguished in this part of the document.

The following measures relevant to spatial planning are proposed, addressing the relevant sectoral planning agencies:

- Natural hazard mitigation should be integrated into all areas relevant to spatial planning (Gobierno Regional de Aysén, 2019, p. 116).
- It is proposed that risk perceptions be studied and taken into account in the action plans to minimize risks from natural hazards (Gobierno Regional de Aysén, 2019, p. 120).
- Erosion control measures (Gobierno Regional de Aysén, 2019, p. 121)
- Promotion of forest management plans to reduce the risk of wildfires (Gobierno Regional de Aysén, 2019, p. 120)

4.4.3 Land Use Zoning

The preferential land uses that exist in the land zoning of the Regional Plan are: forest (protected and preserved), forests with limited suitability and woodlands, agroforestry (silvopasture), livestock, farming (livestock and agriculture), farming with priority agriculture (Fig. 4.3). Each of these categories also exists as a mirror category including tourist use. There is no category for reforestation or afforestation, although it is mentioned in the text part (División de Planificación y Desarrollo Regional, 2013, p. 363).

There is, however, one category of pre-forest use that we highlight and analyze in more detail because of its particular relevance to the issue at hand: Agroforestry. Agroforestry use (silvopastoral system) is defined as agricultural and livestock practices in open forest (open canopy) on degraded soils (División de Planificación y Desarrollo Regional, 2013, p. 307). The geospatial analysis shows that these zones the allocated in the vicinity of forest plantations, which are *Pinus*, and therefore exotic tree species (Fig. 4.3). This is relevant for the conformity assessment, as will be discussed explained in the Sect. 4.4.4.1.

4.4 Results

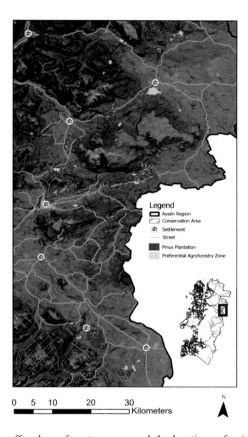

Fig. 4.3 Map of preffered agroforestry zones and the locations of existing *Pinus* plantations. (Selected representation of the geospatial analysis with an enlargement of the situation around the regional capital of Coyhaique. The Pinus plantations present in 2018 are shown in red. In yellow are the zones designated for preferred agroforestry use in the land zoning, extracted from the land zoning (División de Planificación y Desarrollo Regional, 2013, p. 324). According to the corresponding text section of the regional plan, this does not refer to traditional agroforestry, but to afforestation with exotic *Pinus* tree plantations and the planning for compensation payments until the trees are large enough to no longer be damaged by browsing (División de Planificación y Desarrollo Regional, 2015, p. 363). Thus, agroforestry planning category serves a Trojan horse for the expansion of the exotic *Pinus* tree plantations (Cartography: Danny Tröger, Data: Hernández-Moreno et al., 2023, División de Planificación y Desarrollo Regional, ESRI, NASA))

4.4.4 Conformance Evaluation

Conformance Evaluation focuses on the conformity of the land zoning to the goals set in the regional plan.

4.4.4.1 Environmental Goals and Land Use Zoning

There are large areas designated for forest conservation and protection (see purple striped areas in Fig. 4.1 and Fig. 4.2). These areas maintain the resilience of landscapes and thus meet the objective of preventing erosion and mitigating natural hazards.

The woodland category allows for wood extraction in the native forests. If managed effectively, wood extraction per se does not pose a risk to the environmental objectives (Sect. 4.4.1.1), but the impact depends on management practices (McFadden and Dirzo, 2018). This regulatory competence (rule-making, control, and enforcement) lies with the forestry agency CONAF (CONAF, 2015; Di Giminiani, 2016b). However, spatial planning does have an influence on where timber extraction takes place, mainly through the planning of transport infrastructure.

The text section of the regional plan announces measures for preferential afforestation zones, but this category does not exist in the land zoning. Instead, the document refers to agroforestry zones with (exotic) tree plantations (División de Planificación y Desarrollo Regional, 2013, p. 363). Thus, agroforestry zones **do not** refer to the traditional form of animal husbandry in a forest, as practiced for example by some Mapuche communities in the north of the region and on the border with Argentina (Chillo et al., 2021). Instead, the regional plan refers to exotic *Pinus* plantations, where livestock are kept. The spatial location of the areas designated for agroforestry in the vicinity of existing forest plantations indicates that the agroforestry concept serves to facilitate the further expansion of exotic tree plantations in the region (Fig. 4.3).

In summary, land zoning partially supports the environmental objectives of the Regional Plan. The land zoning is inconsistent in its definition and understanding of reforestation, and invasive species management. While the eradication of invasive species is stated as an objective (División de Planificación y Desarrollo Regional, 2013, p. 363), the definition of agroforestry favours invasive *Pinus* species.

4.4.4.2 Identity Goals and Land Zoning

The proposed actions in the Regional Plan to achieve the objectives and goals of identity preservation and equality of opportunity for indigenous communities is to identify the situation of indigenous peoples and incorporate the cultural perspective in all public initiatives (División de Planificación y Desarrollo Regional, 2013, p. 370).

The Regional Plan devotes three pages in the Annex to the fact that different indigenous populations live in the region, some of them as descendants of refugees (División de Planificación y Desarrollo Regional, 2013, pp. 403–406). It formulates objectives goals for the promotion of indigenous peoples. However, there is a lack of concrete instruments and proposals for translating identity objectives into actionable measures.

Land zoning does not include the practice of agroforestry systems with transhumance under native tree species as a category, which is still widespread among indigenous communities (Chillo et al., 2021). There is no evidence of a spatial implementation of the objectives in support of indigenous communities. Therefore land zoning is not in line with the identity objectives.

4.4.5 Performance Evaluation

The performance of the regional plan and land zoning refers to the extent to which they influence relevant decisions, and are implemented in subordinate instruments and policies.

There are no references to the regional plan in the municipal and intermunicipal regulatory plans of the Aysén region, which are intended to substantiate the land zoning (Municipalidad de Coyhaique, 2011). The programs of the Forest Servicey (CONAF) do not correspond with the regional plan; they are not coordinated with the land zoning[1]. The regional biodiversity strategy does not refer to the regional plan (SEREMI MMA-AYSÉN, 2018).

Chile's own spatial tool for supporting indigenous communities in Chile, the *Areas de desarrollo indigena* (Indigenous Development Areas), has not been used at all in the Aysén region to date (División de Planificación y Desarrollo Regional, 2013, p. 170). There is no indication in the Regional Plan, either in the text or on the maps, that the regional government intends to work toward the establishment of such areas.

The overall performance of the PROT and its land zoning for the governance of landscape resilience is very limited. The impact on actual land use is even lower than the influence on land use policy, as the preferred use zones cannot

---

[1] Personal Communication with Local CONAF Staff, September 2023. For reasons of data protection, more detailed information can only be provided on request.

change existing land uses on private land (División de Planificación y Desarrollo Regional, 2013, p. 320).

4.5 Discussion

The aim of the planning evaluation was to determine the extent to which the risks of biodiversity loss, natural hazards and the displacement of indigenous identity are taken into account in spatial planning. A brief description of the shortcomings in this regad is followed by an interpretation in terms of the polycentric governance paradigm. Based on this, suggestions are made on how spatial planning and land use governance can better contribute to landscape resilience.

4.5.1 Low Performing Biodiversity Conservation and Natural Hazard Mitigation

The land use zoning established for the Aysen Region is dedicated to sustainable development (Franchi-Arzola et al., 2018) and uses the new tool of land zoning to designate agroforestry zones with the aim of solving the problem of soil degradation (Ortiz et al., 2023; Ossandón et al., 2020; Peri et al., 2016; Vicencio et al., 2023). Agroforestry (silvopastoral systems) as a land use category is in principle suitable for achieving the environmental objectives of erosion control and also has advantages in terms of protection against natural hazards (Caballé et al., 2016; Ortiz et al., 2023; Peri et al., 2016; Sotomayor et al., 2016; Zinngrebe et al., 2020). However, with agroforestry the regional plan refers to afforestation with exotic tree plantations where livestock is also kept (División de Planificación y Desarrollo Regional, 2013, p. 363). This land use is therefore likely to be counterproductive (Tröger et al., 2022). In order for forage grasses to grow on exotic tree plantations, they have to be sown, which is why European seed mixtures are used (Sotomayor et al., 2016). This contradicts the environmental action plan to presere biodiversity (División de Planificación y Desarrollo Regional, 2013, p. 367) and limit or eradicate invasive species, as stated on page 369 (División de Planificación y Desarrollo Regional, 2013). Some of the species used for tree plantations are clearly invasive, such as *Pinus contorta* (Bravo-Monasterio et al., 2016; Langdon et al., 2010). Other *Pinus* species and the Douglas fir (*Pseudotsuga Menziesii*), also threaten the biodiversity of other plants, insects and birds (Braun et al., 2017b; Paritsis and Aizen, 2008). Forest plantations without sowing of forage grasses are only suitable as pasture for small livestock such as goats, and

4.5 Discussion

only to a limited extent: The needles are difficult to digest, especially those that grow during dry periods due to the increased lignin content (Caballé et al., 2016). These dry periods are likely occur more frequently and last longer due to future climate change, putting the fodder supply for livestock at risk (Olivares-Contreras et al., 2019).

Pinus plantations provide limited erosion control because they do not improve the relevant soil properties (Banfield et al., 2018; Tröger et al., 2022). The protective herb layer is suppressed due to canopy shading and the needle litter layer, and the risk of wildfires is increased, after which large-scale erosion occurs (Bonilla et al., 2014; Braun et al., 2017b; Toro-Manríquez et al., 2019; Tröger et al., 2022). They also provide less landslide mitigation than native species (Tröger et al., 2022).

For agroforestry to reach its full potential as a soil conservation and natural hazard mitigation measure, there needs to be a clear commitment against the use of invasive tree species such as *Pinus* for agroforestry and restoration of eroded areas is needed. Instead, reforestation with native species is possible and has ecological benefits (Zamorano-Elgueta and Moreno, 2021), yet requires more effort and knowledge to establish (Bannister et al., 2018; Huertas Herrera et al., 2022; Kremer et al., 2022). Government support for agroforestry with native tree species should therefore be expanded. Spatial planning and land zoning have their limits here, as this would be a sectoral regulation. Nevertheless, land zoning can be important if, for example, support programs are tied to the location in a preferred zone for agroforestry.

4.5.2 Neglected (Indigenous) Risk Perception

Despite the fact that the Aysén region has the highest percentage of indigenous people in all of Chile, their right to territories it is not recognized (División de Planificación y Desarrollo Regional, 2013, p. 170). This is often justified with the argument, that these are not their ancestral territories and they are not living their ancestral lifestyles (Funk, 2012). Instead, the regional plan foresees the inclusion of indigenous communities in a process of modernization (División de Planificación y Desarrollo Regional, 2013, p. 370). This reiterates the hegemonic paradigm that the land should become productive (Cid-Aguayo et al., 2022; Escalona Ulloa and Barton, 2020; Hamilton and Ramcilovic-Suominen, 2023). There is no reception of their risk perception or cosmovision (Barton et al., 2019; Barton et al., 2012; Huaiquimilla-Guerrero et al., 2022). Transformation of the land surface for extractive purposes, using exotic trees, contradicts their cosmovision and landscape image (Ranjan et al., 2021b; Tröger and Braun). Traditional agroforestry practices with native tree species, on the other hand, would have synergistic effects with ecological resilience mechanisms, as described in Sect. 5.1 (Chillo et al., 2021; Sepúlveda and Guyot, 2016).

The potential of agroforestry as a preferred land use for ecological resilience and to address local, particularly indigenous, risk perceptions is not being realized. Instead, the concept is being used to establish more forest plantations in the region and thereby promoting the further erosion of traditional land uses.

4.5.3 Inspirations from Polycentric Governance

The creation of the regional plan with the land zoning takes place within a larger context of decentralization in Chile. Land use governance in Chile has historically been highly hierarchical (Vicencio et al., 2023). There were central governance mechanisms that affected rural land use—such as the Decree No. 701 on subsidies for tree plantation (Bopp et al., 2020)—but without territorial coordination or spatial planning. Meanwhile, the regional government, is responsible for land use planning for the territory (Vicencio et al., 2023). Thus, it is expected that the planning will be more respectful of the environmental and social conditions of the territory (Ossandón et al., 2020).

The introduction of regional planning with land use zoning above the municipal level is consistent with general findings in the PG literature. Government agencies can pursue a more general mandate at a higher administrative level and have the potential to counteract local inequities, patronage and discrimination—at least in theory (Cox et al., 2010). It could also act as a counterweight to powerful interests at the national administrative level, where local conditions and demands are often neglected (Morrison et al., 2019). In general, the chances of successful regulation are higher, when different levels of the government and the local users themselves are involved in negotiating the rules of use and the system of sanctions (Ibrahim et al., 2023; Ostrom, 1994; Thiel et al., 2015). In theory, regional planning and land zoning could contribute to such conditions. Regional land zoning can be an important link between global, regional and local environmental sustainability goals, ensuring the provision of ecosystem services including natural hazard mitigation and landscape resilience, and climate change adaptation. It can set the stage for resilient land use and define zones of particular importance for natural hazard mitigation. Because of the geographic scale of the planning, it is also an appropiate scale for addressing climate change and nature-based solutions for natural hazard mitigation (Morrison et al., 2022). In particular, agroforestry

4.5 Discussion

zones offer potential for sustainable land use and landscape resilience potentials (Peri et al., 2021). However, there is a pitfall: the rules for which species to use are allowed to be used, would be the task of sectoral planning, which is still hierarchical. This would require consensus on environmental goals, and constant coordination between regional spatial and national, regional and local sectoral planning. One problem that can arise here is that efforts to achieve sustainability in regional planning can be undermined by lobbying at the national level in the sectoral planning (Heilmayr and Lambin, 2016), as the national level is often interested in the export-oriented, economically accountable values (Morales Olmos, 2022).

At the national level, there is a need to coordinate the establishment of sustainable management practices that are regionally specific and adapted to regional climate, vegetation composition and ecological capacities. At the local level, Coordination is needed to drive and manage the implementation of the land zoning requirements in subordinate, regulatory plans at the community level. An important task here is to ensure that social-environmental inequalities are not further exacerbated by the establishment of preferential land use zones.

A common problem in spatial planning is that local elites influence planning decisions in their favor. For example, zones that they owe become land for construction (Chiodelli and Moroni, 2015; Richards, 2010). Since, for historical reasons, people of indigenous origin usually already live in the least ecologically productive zones already and rarely participate in local municipal decisions, regional planning should pay special attention to the interests of these groups (Jullian and Nahuelhual, 2021; Rasmussen and Figueroa, 2022; Sager, 2022). In the present case, this problem of elite capture manifests itself mainly in two phenomena: First, the fact that the traditional agroforestry practices with transhumance of the indigenous population have not been included in the land zoning, but agroforestry is defined as a category of afforestation category with exotic tree species and thus becomes a category for investment. Secondly, the zones for agroforestry zones with exotic species are located near existing *Pinus* plantations, which makes them economically lucrative, but not necessarily useful for erosion control (Fig. 4.3).

Environmental planning in Chile, including spatial planning, shows the emergence of polycentricity (Arriagada et al., 2018; Ossandón et al., 2020; Vicencio et al., 2023). However, to date, some of the key design principles of normative polycentric governance have not been implemented: Adaptation to local socio-environmental conditions is inadequate, as the planning evaluation has shown, and there is no relevant local voice in the land use governance. A widespread proposal in normative polycentric governance to counteract this, is community forest management (Bry et al., 2008; Goetz et al., 2023; Nagendra and Ostrom, 2012; Sosa et al., 2023).

4.5.4 From Critique to Proposals

As discussed in Sects. 4.4.4.1 and 4.5.1, the environmental objectives and the objectives of the regional planning and the land use zoning are partially consistent. Inconsistencies can be observed in land use zoning. On the one hand, the regional plan sets the goal of eradicating invasive species, while on the other, the action plan promotes erosion control through forest plantations with exotic tree species, but disguises it as agroforestry (División de Planificación y Desarrollo Regional, 2013, 363, 369). According to this argument, invasive species will continue to be promoted, even though they do not contribute to environmental goals. Instead, they are areas of investment.

Another point we want to address is the lack of support for indigenous cosmovisions, in which nature cannot be divided into commodities and assets, nor can it be the property of humans (Del Aigo et al., 2022; Ranjan et al., 2021b; Salazar and Riquelme Maulén, 2022). In northern Patagonia there are examples of agroforestry systems, managed by indigenous communities as common pool resources (Chillo et al., 2021; Peri et al., 2016). Based on this critique, we propose the following three strategies to increase landscape resilience:

1. Promote agroforestry with native tree species and ban of exotic forest plantations (Chillo et al., 2021; Plieninger et al., 2020; Zinngrebe et al., 2020)
2. Co-creation of risk-based land zoning (Barton et al., 2020; Greiving, 2002; Puppim de Oliveira, Jose A. and Fra.Paleo, 2016; van Assche et al., 2022)
3. Enable community forest management zones to support common-pool resource management practices and enable alternative cosmovisions (Devisscher et al., 2021; Dressler et al., 2010; Ibrahim et al., 2023)

4.5.4.1 Promoting Agroforestry with Native Tree Species and Banning of Exotic Forest Plantations

Agroforestry is a land use practice with generally high sustainability and resilience potential (Boillat and Bottazzi, 2020; Rocha et al., 2020; van Noordwijk et al., 2018; Zinngrebe et al., 2020). In addition, there are synergies with respect to natural hazard prevention and climate change adaptation (Kumar and Nair, 2011). However, the degree of sustainability depends on at least four factors.

First, a clear commitment to prohibit the use of exotic tree species such as *Pinus* for agroforestry or forest plantations would be mandatory (see Sect. 5.4). Secondly, the extent and intensity of land use remains an open question in the regional plan. Nowhere is it defined what the density of trees or livestock should be in these areas. Third, the areas designated for agroforestry are relatively few and small (Fig. 4.3). They are located in ecotones that are particularly vulnerable to erosion and where afforestation would be beneficial as protection against natural hazards (Tröger et al., 2022). Agroforestry zones would need to cover larger areas to realize their potential for sustainability and resilience. Fourthly, it is unclear whether the farming categories are intended to exclude agroforestry practices (e.g. livestock). Taken together, these issues seriously distort effectiveness and miss the potential for resilient land use.

Regional planning cannot enforce the prohibition of exotic tree species, because this competence is the responsibility of the sectoral planning (CONAF, 2015). However, regional planning can initiate and maintain coordination between the relevant agencies, such as CONAF and INFOR. A clear announcement in the regional planning and the land zoning against the planting of exotic tree species and the use of exotic species in agroforestry is a prerequisite for coordination to be effective in the long term, to achieve a ban of exotic species and thus improve environmental performance.

4.5.4.2 Co-Creation of Risk-Based Land Zoning

Natural hazards policy in Chile has traditionally been dominated by national state interests and neoliberal policies that dominate the disaster risk and recovery management (Sandoval et al., 2014). Local politicies, which could implement disaster risk reduction measures, are rarely committed to long-term visions (Valdivieso and Andersson, 2017; Valdivieso and Públicas, 2019). With the introduction of regional planning, there is an opportunity to move from reaction to prevention. An important step in land use planning towards natural hazard resilient landscapes would be to integrate disaster risk prevention into land use planning. One suggestion we would like to use for this comes from Greiving (2002, p. 275):

It would not be necessary to assign the entire planning area to one of the risk categories (Table 4.2). Where appropriate, these risk-based categories could be overlaid on the existing designated uses and further explained in the text part of the regional plan (Greiving, 2002, p. 274).

The exact design would depend on the type of hazard and land use. The subordinate regulatory plans would have to adopt the zoning or further specify it spatially (Greiving, 2002, p. 275). In addition, sectoral planning would have to

Table 4.2 Categories of Risk Zones according to Greiving (2002, p. 275)

Risk Priority Areas	In these areas, spatial planning decisions must give priority to risk aspects. Arguments related to the risk of natural hazards have the highest weight here. These areas cover two situations: The first case concerns areas that are affected by natural hazards, and entities such as build assets, productive activities and the land users themselves should be protected from them. The second case concerns areas that are particularly suitable as prevention areas for natural hazards, even if they occur in other areas. Examples of the first case would be areas at risk by landslides or flooding. An example of the second would be slopes that act as buffers during heavy rain events to slow surface runoff and prevent downstream flooding. Here, vegetation should be maintained and reforestation encouraged. In addition to ecological suitability, local risk perception must be taken into account when selecting tree species.
Risk Suitability Area	These areas are low-risk and are particularly suitable for projects that must not be at risk from natural hazards. An example would be an industrial facilities that, if damaged or destroyed by a natural hazard, would itself become a source of further hazards, such as chemical spills.

be linked to risk zoning. Its incentives, in particular permits, funding or penalties, could be linked to risk zoning to make it effective (Greiving, 2002, p. 276).

However, the crucial point is not the administrative design, but another: the creation of the categories should be a process of co-creation so that the risk perception of the affected population is integrated (Barton et al., 2020; Puppim de Oliveira, Jose A. and Fra.Paleo, 2016). For example, it is very likely that native tree species would be better accepted than exotic tree plantations (Tröger and Braun). There are empirical examples from Canada that show that local forest knowledge can contribute to building resilience, which leads to the third strategy (van Assche et al., 2022).

4.5.4.3 Community Forest Management Zones

The Regional Plan of Aysén aims to ensure sustainable land use, that is conducive to landscape resilience (División de Planificación y Desarrollo Regional, 2013, p. 8). However, there are fundamental problems as to why spatial planning can have little impact here. These are rooted in the way in which nature is appropriated, as the land is privately owned and must be economically viable, and the state and local communities have no real participation in the development of nature (Atleo and Boron, 2022; Braun, 2021b; Mudombi-Rusinamhodzi and Thiel, 2020). There are no common pool resources in the study area. However,

4.5 Discussion

this is not irreversible, and it would be helpful to prepare spatial planning for this. One possibility would be to establish a land use category with a preference for community forest management (Bixler, 2014; Bry et al., 2008; Sosa et al., 2023). This would serve two purposes. First, to ensure the participation of land users in the governance of the territory concerned, and second, to provide a space for alternative cosmovision to the modern worldview, thus enabling traditional land use practices (Ranjan et al., 2021b). These traditional land use practices include semi-nomad lifestyles (transhumance) and commons.

The possibilities for implementing community forest management are currently limited under the existing overarching rules and the system of property rights. Most of the land in the Aysén region is currently either a protected conservation areas (national or private), which excludes users, or is privately owned (División de Planificación y Desarrollo Regional, 2013). One option would be for communities to buy back land and then work with user groups to establish rules for its use. Examples from Canada show that this approach and the transfer of land use rights to local communities, including spatial organization and zoning, can be successful(van Assche et al., 2022). First Nations communities have been granted self-determination over land use, including economic activities, as well as financial support (Murphy et al., 2020).

In Chile, an increasingly popular approach is the creation of private biodiversity conservation zones (Di Giminiani and Fonck, 2018). Although these can promote ecological sustainability and landscape resilience, they are often associated with the social exclusion of traditional users (Louder and Bosak, 2019). Thus, they are not an alternative to constitutionally guaranteed common pool resources, as they are highly dependent on the will of the investors (Louder and Bosak, 2019). This telecoupling can provoke conflict with the local communities and can end up in feeding conspiracy theories (Louder and Bosak, 2019).

Another approach would be to use an existing instrument in Chile that has not yet been used in the Aysén region: the Areas for Indigenous Development (AID) (Hiriart-Bertrand et al., 2020; Rojo-Mendoza et al., 2022). The proposed new category of community forest management could be complementary to the areas of indigenous development areas. However, Community Forest Management Zones should go well beyond the purposes and possibilities of AIDs, as they continue to follow the guiding principle of making land economically usable (Ranjan et al., 2021b).

A condition of the community forest management zones should be that the land users are actually the local population and that telecoupling effects should be limited (Boillat et al., 2018; Carrasco et al., 2017; Challies et al., 2019).

They should be designed in a way to guarantee the support of local communities, respecting their landscape image, territorial identity and cosmovision (Braun, 2021b; Chiodelli and Moroni, 2015; Richards, 2010). It is of utmost importance that a different ontology of nature and the environment underlies this (Braun, 2021b). The natural environment is not divisible and interchangeable, cannot be the property of humans and landscape elements are not just objects, but have an emotional-spiritual meaning that is interwoven with human actions (Di Giminiani and González Gálvez, 2018; Schmalz et al., 2023). If this fundamental level is not taken into account, conflicts is inevitable (Braun, 2021b; Schmalz et al., 2023).

4.6 Summary and Conclusions

The question of the extent to which spatial planning, through the instrument of land zoning is suitable for mitigating land-use related risks to landscape resilience was investigated. Because of their particularly important role for landscape resilience, the focus was on temperate forests in the last of the wild areas and subsequent uses, including afforestation. In the study area, the Chilean Patagonian region of Aysén, regional spatial planning and land zoning is dedicated to sustainable land use. However, evaluation of planning can show that the contribution to landscape resilience is limited, if not counterproductive.

Land zoning is not in line with the environmental and identity goals of the regional plan. This is particularly evident in the fact that the resilience potentials that agroforestry practices could offer cannot unfold, as this land use category is (mis-)used to promote the establishment of further forest plantations with exotic species. The ecological processes and functions of forests that promote landscape resilience cannot be developed by such forest plantations and contradict local risk perceptions (Tröger et al., 2022; Tröger and Braun 2024).

The example of Aysén shows why caution is needed when introducing planning instruments, and how land zoning can easily be undermined, threatening the resilience of the landscape and taking on a neo-colonial character. Land use concepts can be reinterpreted when they become planning categories, and reproduce extractive land uses. In this case, agroforestry becomes a Trojan horse for commercial forest plantations – with few resilience-building effects and creating new risks – instead of supporting a traditional land use practice – with many resilience-building effects (Chillo et al., 2021; Streck, 2020; Zinngrebe et al., 2020).

4.6 Summary and Conclusions

The fundamental risks to landscape resilience also exist elsewhere: in the context of a warming climate, the cool temperate and boreal zones of the world's last wilderness areas are also likely to face the question of how to maintain local and regional landscape resilience in the face of increasing deforestation. The strategies chosen are not only relevant for local communities, as forests and soils in these areas may or may not be carbon sinks, depending on how they are managed (Anderies et al., 2013; IPCC, 2023; Thurner et al., 2014). In particular, land use management must take into account biogeomorphic feedbacks of the chosen tree species, which have a decisive influence on natural hazards such as landslides or wildfires (Tröger et al., 2022; Vesterdal et al., 2013). Local perceptions of risk should be prioritized in land use and zoning, as local forest knowledge can greatly enhance resilience (Puppim de Oliveira, Jose A. and Fra.Paleo, 2016; van Assche et al., 2022).

5 What Can We Learn from Aysén?—Discussion and Derivation of Follow-up Research Questions

In this section, the findings from the Chaps. 2, 3 and 4 relevant for the overarching research questions are taken up again. Where appropriate, the conditions for transferability to other study areas are discussed, including the identification of further research gaps.

5.1 Biogeomorphic Feedbacks, Land Use and Landscape Resilience—Underestimated Interactions?

The biogeomorphology study (Chap. 2) showed that vegetation assemblage and land use can significantly change soil properties within a few decades. These include soil parameters associated with susceptibility to erosion processes, landslides and the water balance. Changes in soil properties are strongly dependent on the species present on the site. For example, *Pinus* plantations have positive effect on some soil properties, but increase the risk of wildfires. Modeling can help to scale up these changes in soil properties from qualitative single point measurements to the landscape scale (Gerke et al., 2022; Pohl et al., 2022; Späth et al., 2023; Weber et al., 2022). The importance for landscape resilience can then be quantified. The concept presented in Sect. 2.5 and visualized in Fig. 2.7 helps to connect modeling results from different disciplines that are not (yet) considered together. This is important for meaningful planning.

Considering the multidimensional space of the soil properties (Fig. 2.6), it appears that a location between the points of primary forest and pasture would be favorable for landscape resilience. One such land use that would correspond to this is agroforestry. Indeed, it is plausible that trees make arable and pasture land more resilient (Plieninger et al., 2020). However, long-term studies on experimental plots would be needed to determine the contribution of different forms of agroforestry to the resilience of the sites and their surrounding land uses. Agroforestry is still rare in Germany, so there is a lack of suitable study sites (Hübner et al., 2020). The results of the present work provide an argument for the design of corresponding experiments and studies.

5.2 Risk Perception: Does the Distinction Between Natural Hazards and Environmental Degradation Still Make Sense? Are We Looking at the Wrong Side of Risk?

The study on risk perception (Chap. 3) revealed the following findings: A wide variety of disciplines with different approaches and perspectives have contributed to the research on disaster risk perception. This is reflected in the concepts of emergent order. However, they all followed the separation of natural hazards and environmental degradation, and of hazard and vulnerability. As such, they manifest the nature-culture divide of modernity (Katanha and Simatele, 2019; Latour, 1995). The concepts of a hybrid understanding promise a way out of these dichotomies. They emphasize the need for a more complex understanding of vulnerability. This perspective could be relevant not only for Chile, but also for Germany.

An increasingly large proportion of the ecological processes that constitute natural hazards are anthropogenically influenced (Gill and Malamud, 2017). Leaving aside tectonic processes and volcanic activity, what is the point of distinguishing between natural hazards and environmental problems in the context of risk governance? It can be argued that attribution research can answer the question of the proportions of anthropogenic and natural climate change with increasing precision using sophisticated statistical methods (Kreienkamp et al., 2021). But it remains a stochastic variable. The discourse on it distracts from the question of how to deal with social and socio-spatial vulnerability, which is the more important question for spatial planning. More importantly, it could fuel socio-spatial conflicts related to disaster risk. For example, during the Ahr valley disaster, conspiracy theorists infiltrated the relief effort, using the demarcation between natural to anthropogenic causes of climate change for their purposes (Kühne et al., 2021). Concepts that focus on the perception of (social) vulnerability could help to overcome this conflict line. However, this is a hypothesis that requires further research.

If the definition of disaster risk is the intersection of hazard and vulnerability (Wisner et al., 2010), why then do empirical studies of risk perception rarely assess aspects of vulnerability, but rather focus on hazard perception? Maybe it is time to explore (social) vulnerability perception. Further empirical work is needed to extend the concepts presented in Chap. 3 and to explore their potential for other areas of study.

5.3 Reflection on the Contribution of Land Zoning to Landscape Resilience and Governance Analysis Research Proposal

The contribution of spatial planning and tools such as land zoning for natural hazard mitigation and landscape resilience is limited: Some levers are not the responsibility of spatial planning, but of sectoral planning (Berčák et al., 2023; Greiving et al., 2006). For land uses like forestry, afforestation, agroforestry, or pastures, a more direct approach is to develop the potential for landscape resilience through regulations such as permitted species, soil and forest management (Berčák et al., 2023; Glæsner et al., 2014). However, spatial planners should be aware of the biogeomorphic interactions of land uses and the potential consequences for resilience at the landscape scale, and legislation should be changed to promote incorporation of impacts as part of environmental impact assessments and similar tools.

A comprehensive analysis of land use governance is needed to determine what responsibilities and regulatory options exist beyond spatial planning. The starting point for the analysis must be the ecological impact pathways identified in Chap. 2. The Institutional Analysis and Development (IAD) framework could be useful to identify the relevant actors and leverage points, as this approach takes into account the coupling of social and ecological systems and includes more than just the legal regulations (Ostrom, 1994; Schütze and Thiel, in review; Thiel and Moser, 2019).

Recent extensions of the IAD framework and performance analysis of governance systems include narratives and other epistemic aspects, that can be considered highly relevant to landscape resilience issues (Schütze and Thiel, in review). For example, agroforestry practices, have a high potential for landscape resilience, but are barely conceivable in Germany due to the lack of epistemics and narratives about them (Hübner et al., 2020; Mosquera-Losada et al., 2023; Wilson and Lovell, 2016). Transformative approaches must therefore also consider the narratives.

Notwithoutstanding the constraints and limitations mentioned above, natural hazard protection and landscape resilience should nevertheless be a regular issue in spatial planning at all administrative levels. In Germany, the issue of natural hazards is still absent from public information on spatial planning at the state and regional level in some federal states, like Hessen, whilst on the municipal level, some cities are implementing advanced assessment and mitigation approaches (Greiving et al., 2023; HMWEVW, 2023; Kammerbauer, 2014).

5.4 Lessons Learned from Chilean Patagonia

Even apparently small changes in biogeomorphic interactions due to the land use can limit ecological functions important for landscape resilience. Reduced biodiversity due to monofunctional land use can lead to reduced soil performance in terms of protection against erosion and landslides. It can also increase the risk of forest fires. Asessing these can be methodologically challenging, and requires a combination of field methods, laboratory methods and modelling, as well as approaches that bring together insights from different ecological sub-disciplines to address a problem. The risk perception of laypeople based on traditional ecological knowledge can be a good intuition here. The risk perception study also revealed another aspect: it is not so much the "natural" hazards that cause concern, but rather the extractive land uses by other user groups, combined with the exclusion of the local population from land use governance. It is therefore much more a question of social vulnerability.

The biogeomorphic interactions considered in this thesis bring into focus the neglected effects of land use on neighboring land uses and user groups. They illustrate the need for more complexity than the current widespread homogenized land uses. Agroforestry has great potential, as the Aysen region shows us. However, caution is needed in how land use is legally defined in spatial planning to avoid misuse and counterproductive effects. Otherwise, there is a risk that inappropriate land use will continue. A crucial measure to prevent this is the participation of local user groups. In particular, researchers should particularly focus on vulnerable and marginalized groups. If individual functions of land use are pushed driven to maximum performance, but risks are only shifted as a result, nothing is gained.

References

Abid, M., Schilling, J., Scheffran, J., Zulfiqar, F., 2016. Climate change vulnerability, adaptation and risk perceptions at farm level in Punjab, Pakistan. The Science of the total environment 547, 447–460. https://doi.org/10.1016/j.scitotenv.2015.11.125.

Aburto, F., Cartes, E., Mardones, O., Rubilar, R., 2020. Hillslope soil erosion and mobility in pine plantations and native deciduous forest in the coastal range of south-Central Chile. Land Degradation & Development. https://doi.org/10.1002/ldr.3700.

Adomah Bempah, S., Olav Øyhus, A., 2017. The role of social perception in disaster risk reduction: Beliefs, perception, and attitudes regarding flood disasters in communities along the Volta River, Ghana. International Journal of Disaster Risk Reduction 23, 104–108. https://doi.org/10.1016/j.ijdrr.2017.04.009.

Aguiar, S., Mastrángelo, M.E., Brancalion, P.H., Meli, P., 2021. Transformative governance for linking forest and landscape restoration to human well-being in Latin America. Ecosystems and People 17, 523–538. https://doi.org/10.1080/26395916.2021.1976838.

Aguilea, O.E.F., Tonko, J.P., 2013. Relatos de Viaje Kawésqar: Nómadas Canoeros de la Patagonia Occidental. Ofqui Editores, Temuco, Santiago de Chile.

Alaoui, A., Rogger, M., Peth, S., Blöschl, G., 2018. Does soil compaction increase floods? A review. Journal of Hydrology 557, 631–642. https://doi.org/10.1016/j.jhydrol.2017.12.052.

Albers, C., 2012. Coberturas SIG para la enseñanza de la Geografía en Chile. Universidad de La Frontera, Temuco, Chile.

Alexander, E.R. (Ed.), 2006. Evaluation in Planning: Evolution and Prospects (Urban and regional planning and development series). Routledge, 1 online resource.

Alexander, E.R., 2011. Evaluating planning: what is successful planning and (how) can we measure it?, in: Hull, A., Alexander, E.R., Khakee, A., Woltjer, J. (Eds.), Evaluation for participation and sustainability in planning. Routledge, Abingdon, Oxon, pp. 32–46.

Alexander, E.R., 2016. There is no planning—only planning practices: Notes for spatial planning theories. Planning Theory 15, 91–103. https://doi.org/10.1177/1473095215594617.

Ali, T., Buergelt, P.T., Paton, D., Smith, J.A., Maypilama, E.L., Yuŋgirrŋa, D., Dhamarrandji, S., Gundjarranbuy, R., 2021. Facilitating Sustainable Disaster Risk Reduction in Indigenous Communities: Reviving Indigenous Worldviews, Knowledge, and Practices through

Two-Way Partnering. International journal of environmental research and public health 18. https://doi.org/10.3390/ijerph18030855.

Allan, J.R., Venter, O., Watson, J.E.M., 2017. Temporally inter-comparable maps of terrestrial wilderness and the Last of the Wild. Sci Data 4, 170–187. https://doi.org/10.1038/sdata.2017.187.

Allen, R., Platt, K., Wiser, S., 1995. Platt, K., Wiser, S., 1995. Biodiversity in New Zealand plantations. New Zealand Forestry 39, 26–29.

Allmendinger, P., Tewdwr-Jones, M., 1997. Mind the Gap: Planning Theory–Practice and the Translation of Knowledge into Action—A Comment on Alexander (1997). Environ Plann B Plann Des 24, 802–806. https://doi.org/10.1068/b240802.

Alonso, M.F., Wentzel, H., Schmidt, A., Balocchi, O., 2020. Plant community shifts along tree canopy cover gradients in grazed Patagonian Nothofagus antarctica forests and grasslands. Agroforest Syst 94, 651–661. https://doi.org/10.1007/s10457-019-00427-y.

Amigo, A., 2013. Estimation of tephra-fall and lahar hazards at Hudson Volcano, southern Chile: Insights from numerical models, in: Geological Society of America (Ed.), Understanding Open-Vent Volcanism and Related Hazards, vol. 498. Geological Society of America, pp. 177–199.

Amundson, R., Heimsath, A., Owen, J., Yoo, K., Dietrich, W.E., 2015. Hillslope soils and vegetation. Geomorphology 234, 122–132. https://doi.org/10.1016/j.geomorph.2014.12.031.

Anacona, P.I., Mackintosh, A., Norton, K., 2015. Reconstruction of a glacial lake outburst flood (GLOF) in the Engaño Valley, Chilean Patagonia: Lessons for GLOF risk management. The Science of the total environment 527–528, 1–11. https://doi.org/10.1016/j.scitotenv.2015.04.096.

Anderies, J.M., Carpenter, S.R., Steffen, W., Rockström, J., 2013. The topology of non-linear global carbon dynamics: from tipping points to planetary boundaries. Environ. Res. Lett. 8, 44048. https://doi.org/10.1088/1748-9326/8/4/044048.

Anderson, B., McFarlane, C., 2011. Assemblage and geography. Area 43, 124–127. https://doi.org/10.1111/j.1475-4762.2011.01004.x.

Antronico, L., Pascale, F. de, Coscarelli, R., Gullà, G., 2020. Landslide risk perception, social vulnerability and community resilience: The case study of Maierato (Calabria, southern Italy). International Journal of Disaster Risk Reduction 46, 101529. https://doi.org/10.1016/j.ijdrr.2020.101529.

Araujo, P.I., Austin, A.T., 2020. Exotic pine forestation shifts carbon accumulation to litter detritus and wood along a broad precipitation gradient in Patagonia, Argentina. Forest Ecology and Management 460, 117902. https://doi.org/10.1016/j.foreco.2020.117902.

Armesto, J.J., Manuschevich, D., Mora, A., Smith-Ramirez, C., Rozzi, R., Abarzúa, A.M., Marquet, P.A., 2010. From the Holocene to the Anthropocene: A historical framework for land cover change in southwestern South America in the past 15,000 years. Land Use Policy 27, 148–160. https://doi.org/10.1016/j.landusepol.2009.07.006.

Armesto, J.J., Rozzi, R., Smith-Ramírez, C., Arroyo, M.T.K., 1998. Conservation Targets in South American Temperate Forests. Science 282, 1271–1272. https://doi.org/10.1126/science.282.5392.1271.

Arora, S., Stirling, A., 2023. Colonial modernity and sustainability transitions: A conceptualisation in six dimensions. Environmental Innovation and Societal Transitions 48, 100733. https://doi.org/10.1016/j.eist.2023.100733.

References

Arriagada, R., Aldunce, P., Blanco, G., Ibarra, C., Moraga, P., Nahuelhual, L., O'Ryan, R., Urquiza, A., Gallardo, L., 2018. Climate change governance in the anthropocene: emergence of polycentrism in Chile. Elementa: Science of the Anthropocene 6, 68. https://doi.org/10.1525/elementa.329.

Asmüssen, M.V., Simonetti, J.A., 2007. Can a developing country like Chile invest in biodiversity conservation? Envir. Conserv. 34. https://doi.org/10.1017/S0376892907004183.

Astaburuaga, J., Leszczynski, A., Martin, M.E., Gaillard, J.C., 2023. The multiple environmentalities of conservation mapping in Patagonia-Aysén. Environment and Planning E: Nature and Space 6, 1940–1965. https://doi.org/10.1177/25148486221125228.

Atauri, J.A., Pablo, C.L. de, Agar, P.M. de, Schmitz, M.F., Pineda, F.D., 2004. Effects of management on understory diversity in the forest ecosystems of northern Spain. Environmental Management 34, 819–828. https://doi.org/10.1007/s00267-004-0180-0.

Atleo, C., Boron, J., 2022. Land Is Life: Indigenous Relationships to Territory and Navigating Settler Colonial Property Regimes in Canada. Land 11, 609. https://doi.org/10.3390/land11050609.

Azócar García, G., Aguayo Arias, M., Henríquez Ruiz, C., Vega Montero, C., Sanhueza Contreras, R., 2010. Patrones de crecimiento urbano en la Patagonia chilena: el caso de la ciudad de Coyhaique. Rev. geogr. Norte Gd. https://doi.org/10.4067/S0718-34022010000200005.

Bachmann-Vargas, P., van Koppen, C., 2020. Disentangling Environmental and Development Discourses in a Peripheral Spatial Context: The Case of the Aysén Region, Patagonia, Chile. The Journal of Environment & Development 29, 366–390. https://doi.org/10.1177/1070496520937041.

Baldwin, E., Thiel, A., McGinnis, M., Kellner, E., 2023. Empirical research on polycentric governance: Critical gaps and a framework for studying long-term change. Policy Studies Journal, psj.12518. https://doi.org/10.1111/psj.12518.

Balzano, B., Tarantino, A., Ridley, A., 2019. Preliminary analysis on the impacts of the rhizosphere on occurrence of rainfall-induced shallow landslides. Landslides 16, 1885–1901. https://doi.org/10.1007/s10346-019-01197-5.

Banerjee, S.B., Arjaliès, D.-L., 2021. Celebrating the End of Enlightenment: Organization Theory in the Age of the Anthropocene and Gaia (and why neither is the solution to our ecological crisis). Organization Theory 2, 263178772110367. https://doi.org/10.1177/26317877211036714.

Banfield, C.C., Braun, A.C., Barra, R., Castillo, A., Vogt, J., 2018. Erosion proxies in an exotic tree plantation question the appropriate land use in Central Chile. CATENA 161, 77–84. https://doi.org/10.1016/j.catena.2017.10.017.

Bannister, J.R., Vargas-Gaete, R., Ovalle, J.F., Acevedo, M., Fuentes-Ramirez, A., Donoso, P.J., Promis, A., Smith-Ramírez, C., 2018. Major bottlenecks for the restoration of natural forests in Chile. Restor Ecol 26, 1039–1044. https://doi.org/10.1111/rec.12880.

Barbosa, O., Marquet, P.A., 2002. Effects of forest fragmentation on the beetle assemblage at the relict forest of Fray Jorge, Chile. Oecologia 132, 296–306. https://doi.org/10.1007/s00442-002-0951-3.

Barroetaveña, C., Rajchenberg, M., Cázares, E., 2005. Mycorrhizal fungi in Pinus ponderosa introduced in Central Patagonia (Argentina). nova hedw. 80, 453–464. https://doi.org/10.1127/0029-5035/2005/0080-0453.

Barton, J., Román, Á., Fløysand, A., 2012. Resource Extraction and Local Justice in Chile: Conflicts Over the Commodification of Spaces and the Sustainable Development of Places, in: New Political Spaces in Latin American Natural Resource Governance. Palgrave Macmillan, New York, pp. 107–128.

Barton, J.R., Fløysand, A., 2010. The political ecology of Chilean salmon aquaculture, 1982–2010: A trajectory from economic development to global sustainability. Global Environmental Change 20, 739–752. https://doi.org/10.1016/j.gloenvcha.2010.04.001.

Barton, J.R., Román, Á., 2016. Sustainable development? Salmon aquaculture and late modernity in the archipelago of Chiloé, Chile. Island Studies Journal 11, 651–672.

Barton, J.R., Román, Á., Rehner, J., 2019. Responsible research and innovation (RRI) in Chile: from a neostructural productivist imperative to sustainable regional development? European Planning Studies 27, 2510–2532. https://doi.org/10.1080/09654313.2019.1658719.

Barton, T.M., Beaven, S.J., Cradock-Henry, N.A., Wilson, T.M., 2020. Knowledge sharing in interdisciplinary disaster risk management initiatives: cocreation insights and experience from New Zealand. 1708–3087 25. https://doi.org/10.5751/ES-11928-250425

Basile, A., Mele, G., Terribile, F., 2003. Soil hydraulic behaviour of a selected benchmark soil involved in the landslide of Sarno 1998. Geoderma 117, 331–346. https://doi.org/10.1016/S0016-7061(03)00132-0.

Baum, K.A., Haynes, K.J., Dillemuth, F.P., Cronin, J.T., 2004. The Matrix Enhances the Effectiveness of Corridors and Stepping Stones. Ecology 85, 2671–2676. https://doi.org/10.1890/04-0500.

Bavikatte, K.S., Bennett, T., 2015. Community stewardship: the foundation of biocultural rights. JHRE 6, 7–29. https://doi.org/10.4337/jhre.2015.01.01.

Belardi, J.B., 2021. Ancient Hunting Strategies in Southern South America. Springer International Publishing, Cham, 1 online resource.

Beller, E., Robinson, A., Grossinger, R., 2015. Landscape resilience framework: Operationalizing ecological resilience at the landscape scale. San Francisco Estuary Institute, Aquatic Science Center, [San Francisco, CA], 31 pp.

Bendito, A., 2020. Grounding urban resilience through transdisciplinary risk mapping. Urban Transform 2, 1–11. https://doi.org/10.1186/s42854-019-0005-3.

Bennett, M.M., Chen, J.K., Alvarez León, L.F., Gleason, C.J., 2022. The politics of pixels: A review and agenda for critical remote sensing. Progress in Human Geography 46, 729–752. https://doi.org/10.1177/03091325221074691.

Berčák, R., Holuša, J., Kaczmarowski, J., Tyburski, Ł., Szczygieł, R., Held, A., Vacik, H., Slivinský, J., Chromek, I., 2023. Fire Protection Principles and Recommendations in Disturbed Forest Areas in Central Europe: A Review. Fire 6, 310. https://doi.org/10.3390/fire6080310.

Berkes, F., Colding, J., Folke, C., 2006. Navigating social-ecological systems: Building resilience for complexity and change. Cambridge Univ. Press, Cambridge, 393 pp.

Berndt, L.A., Brockerhoff, E.G., Jactel, H., 2008. Relevance of exotic pine plantations as a surrogate habitat for ground beetles (Carabidae) where native forest is rare. Biodiversity and Conservation 17, 1171–1185. https://doi.org/10.1007/s10531-008-9379-3.

Biggs, R. (Ed.), 2015. Principles for Building Resilience. Cambridge University Press.

References

Biggs, R., Schlüter, M., Biggs, D., Bohensky, E.L., BurnSilver, S., Cundill, G., Dakos, V., Daw, T.M., Evans, L.S., Kotschy, K., Leitch, A.M., Meek, C., Quinlan, A., Raudsepp-Hearne, C., Robards, M.D., Schoon, M.L., Schultz, L., West, P.C., 2012. Toward Principles for Enhancing the Resilience of Ecosystem Services. Annu. Rev. Environ. Resour. 37, 421–448. https://doi.org/10.1146/annurev-environ-051211-123836.

Birkmann, J., 2007. Risk and vulnerability indicators at different scales: Applicability, usefulness and policy implications. Environmental Hazards 7, 20–31. https://doi.org/10.1016/j.envhaz.2007.04.002.

Birkmann, J., Cardona, O., Carreño, M., Barbat, A., Pelling, M., Schneiderbauer, Kienberger, S., Keiler, M., Alexander, D., Zeil, P., Welle, T., 2013. Framing vulnerability, risk and societal responses: the MOVE framework. Natural Hazards.

Birkmann, J., Wenzel, F., Greiving, S., Garschagen, M., Vallée, D., Nowak, W., Welle, T., Fina, S., Goris, A., Rilling, B., Fiedrich, F., Fekete, A., Cutter, S.L., Düzgün, S., Ley, A., Friedrich, M., Kuhlmann, U., Novák, B., Wieprecht, S., Riegel, C., Thieken, A., Rhyner, J., Ulbrich, U., Mitchell, J.K., 2016. Extreme Events, Critical Infrastructures, Human Vulnerability and Strategic Planning: Emerging Research Issues. J. of Extr. Even. 03, 1650017. https://doi.org/10.1142/S2345737616500172.

Bixler, R.P., 2014. From Community Forest Management to Polycentric Governance: Assessing Evidence from the Bottom Up. Society & Natural Resources 27, 155–169. https://doi.org/10.1080/08941920.2013.840021.

Bizama, G., Torrejón, F., Aguayo, M., Muñoz, M.D., Echeverría, C., Urrutia, R., 2011. Pérdida y fragmentación del bosque nativo en la cuenca del río Aysén (Patagonia-Chile) durante el siglo XX. RGNG, 125–138. https://doi.org/10.4067/S0718-34022011000200008.

Blackhall, M., Raffaele, E., 2019. Flammability of Patagonian invaders and natives: When exotic plant species affect live fine fuel ignitability in wildland-urban interfaces. Landscape and Urban Planning 189, 1–10. https://doi.org/10.1016/j.landurbplan.2019.04.002.

Bohle, H.-G., 2008. Leben mit Risiko - Resilience als neues Paradigma für die Risikowelt von morgen, in: Felgentreff, C., Glade, T. (Eds.), Naturrisiken und Sozialkatastrophen. Spektrum Akad. Verl., Berlin.

Bohnsack, R., 2014. Rekonstruktive Sozialforschung: Einführung in qualitative Methoden, 9th ed. Budrich, Opladen, 316 pp.

Bohnsack, R., Nentwig-Gesemann, I., 2011. Typenbildung, in: Bohnsack, R., Marotzki, W., Meuser, M. (Eds.), Hauptbegriffe qualitativer Sozialforschung, 3rd ed. Verlag Barbara Budrich, Opladen, Farmington Hills, MI, pp. 162–166.

Boillat, S., Bottazzi, P., 2020. Agroecology as a pathway to resilience justice: peasant movements and collective action in the Niayes coastal region of Senegal. International Journal of Sustainable Development & World Ecology 27, 662–677. https://doi.org/10.1080/13504509.2020.1758972.

Boillat, S., Gerber, J.-D., Oberlack, C., Zaehringer, J., Ifejika Speranza, C., Rist, S., 2018. Distant Interactions, Power, and Environmental Justice in Protected Area Governance: A Telecoupling Perspective. Sustainability 10, 3954. https://doi.org/10.3390/su10113954.

Boillat, S., Martin, A., Adams, T., Daniel, D., Llopis, J., Zepharovich, E., Oberlack, C., Sonderegger, G., Bottazzi, P., Corbera, E., Ifejika Speranza, C., Pascual, U., 2020. Why telecoupling research needs to account for environmental justice. Journal of Land Use Science 15, 1–10. https://doi.org/10.1080/1747423X.2020.1737257.

Bonilla, C.A., Pastén, P.A., Pizarro, G.E., González, V.I., Carkovic, A.B., Céspedes, R.A., 2014. Forest Fires and Soil Erosion Effects on Soil Organic Carbon in the Serrano River Basin (Chilean Patagonia), in: Hartemink, A.E., McSweeney, K. (Eds.), Soil Carbon. Springer International Publishing, Cham, pp. 229–239.

Bonnesoeur, V., Locatelli, B., Guariguata, M.R., Ochoa-Tocachi, B.F., Vanacker, V., Mao, Z., Stokes, A., Mathez-Stiefel, S.-L., 2019. Impacts of forests and forestation on hydrological services in the Andes: A systematic review. Forest Ecology and Management 433, 569–584. https://doi.org/10.1016/j.foreco.2018.11.033.

Bopp, C., Engler, A., Jara-Rojas, R., Arriagada, R., 2020. Are forest plantation subsidies affecting land use change and off-farm income? A farm-level analysis of Chilean small forest landowners. Land Use Policy 91, 104308. https://doi.org/10.1016/j.landusepol.2019.104308.

Boyle, J.R., Powers, R.F., 2013. Forest Soils, in: Reference Module in Earth Systems and Environmental Sciences. Elsevier.

Braun, A.C., (in Review). Deforestation caused by afforestation: A spectral-spatial classification framework applied to the case of Central Chile. Remote Sensing.

Braun, A.C., 2021a. Encroached by pine and eucalyptus? A grounded theory on an environmental conflict between forest industry and smallholder livelihoods in Chile. Journal of Rural Studies 82, 107–120. https://doi.org/10.1016/j.jrurstud.2021.01.029.

Braun, A.C., 2021b. Encroached by pine and eucalyptus? A grounded theory on an environmental conflict between forest industry and smallholder livelihoods in Chile. Journal of Rural Studies 82, 107–120. https://doi.org/10.1016/j.jrurstud.2021.01.029.

Braun, A.C., 2021c. More accurate less meaningful? A critical physical geographer's reflection on interpreting remote sensing land-use analyses. Progress in Physical Geography: Earth and Environment 45, 706–735. https://doi.org/10.1177/0309133321991814.

Braun, A.C., 2022. Deforestation by Afforestation: Land Use Change in the Coastal Range of Chile. Remote Sensing 14, 1686. https://doi.org/10.3390/rs14071686.

Braun, A.C., Faßnacht, F., Valencia, D., Sepulveda, M., 2021. Consequences of land-use change and the wildfire disaster of 2017 for the central Chilean biodiversity hotspot. Reg Environ Change 21, 1–22. https://doi.org/10.1007/s10113-021-01756-4.

Braun, A.C., Rojas, C., Echeverri, C., Rottensteiner, F., Bahr, H.-P., Niemeyer, J., Arias, M.A., Kosov, S., Hinz, S., Weidner, U., 2014. Design of a Spectral–Spatial Pattern Recognition Framework for Risk Assessments Using Landsat Data—A Case Study in Chile. IEEE Journal of Selected Topics in Applied Earth Observations and Remote Sensing 7, 917–928. https://doi.org/10.1109/JSTARS.2013.2293421.

Braun, A.C., Troeger, D., Garcia, R., Aguayo, M., Barra, R., Vogt, J., 2017a. Assessing the impact of plantation forestry on plant biodiversity. Global Ecology and Conservation 10, 159–172. https://doi.org/10.1016/j.gecco.2017.03.006.

Braun, A.C., Troeger, D., Garcia, R., Aguayo, M., Barra, R., Vogt, J., 2017b. Erosion proxies in an exotic tree plantation question the appropriate land use in Central Chile. Global Ecology and Conservation 10, 159–172. https://doi.org/10.1016/j.gecco.2017.03.006.

Braun, A.C., Vogt, J., 2014. A Multiscale Assessment of the Risks Imposed by Plantation Forestry on Plant Biodiversity in the Hotspot Central Chile. OJE 04, 1025–1044. https://doi.org/10.4236/oje.2014.416085.

Braun-Blanquet, J., 1964a. Pflanzensoziologie. Grundzüge der Vegetationskunde. Springer, Wien.

Braun-Blanquet, J., 1964b. Pflanzensoziologie. Grundzüge der Vegetationskunde. Springer, Wien.
Bravo-Monasterio, P., Pauchard, A., Fajardo, A., 2016. Pinus contorta invasion into treeless steppe reduces species richness and alters species traits of the local community. Biol Invasions 18, 1883–1894. https://doi.org/10.1007/s10530-016-1131-4.
Bremer, L.L., Farley, K.A., 2010. Does plantation forestry restore biodiversity or create green deserts? A synthesis of the effects of land-use transitions on plant species richness. Biodivers Conserv 19, 3893–3915. https://doi.org/10.1007/s10531-010-9936-4.
Breuer, F., Muckel, P., Dieris, B., Allmers, A., 2019. Reflexive Grounded Theory: Eine Einführung für die Forschungspraxis, 4th ed. Springer VS, Wiesbaden, 439 pp.
Brockerhoff, E.G., Berndt, L.A., Jactel, H., 2005. Role of exotic pine forests in the conservation of the critically endangered New Zealand ground beetle Holcaspis brevicula (Coleoptera: Carabidae). New Zealand Journal of Ecology 29, 37–43.
Brockerhoff, E.G., Ecroyd, C.E., Leckie, A.C., Kimberley, M.O., 2003. Diversity and succession of adventive and indigenous vascular understorey plants in Pinus radiata plantation forests in New Zealand. Forest Ecology and Management 185, 307–326. https://doi.org/10.1016/S0378-1127(03)00227-5.
Brockerhoff, E.G., Jactel, H., Parrotta, J.A., Quine, C.P., Sayer, J., 2008. Plantation forests and biodiversity: oxymoron or opportunity? Biodivers Conserv 17, 925–951. https://doi.org/10.1007/s10531-008-9380-x.
Brooks, T.M., Mittermeier, R.A., Da Fonseca, G.A.B., Gerlach, J., Hoffmann, M., Lamoreux, J.F., Mittermeier, C.G., Pilgrim, J.D., Rodrigues, A.S.L., 2006. Global biodiversity conservation priorities. Science (New York, N.Y.) 313, 58–61. https://doi.org/10.1126/science.1127609.
Bry, D.B., Duran, E., Ramos, V.H., Mas, J.-F., Velázquez, A., McNab, R.B., Barry, D., Radachowsky, J., 2008. Tropical Deforestation, Community Forests, and Protected Areas in the Maya Forest. Ecology and Society 13.
Buma, B., Harvey, B.J., Gavin, D.G., Kelly, R., Loboda, T., McNeil, B.E., Marlon, J.R., Meddens, A.J.H., Morris, J.L., Raffa, K.F., Shuman, B., Smithwick, E.A.H., McLauchlan, K.K., 2019. The value of linking paleoecological and neoecological perspectives to understand spatially-explicit ecosystem resilience. Landscape Ecol 34, 17–33. https://doi.org/10.1007/s10980-018-0754-5.
Buschmann, A.H., Cabello, F., Young, K., Carvajal, J., Varela, D.A., Henríquez, L., 2009. Salmon aquaculture and coastal ecosystem health in Chile: Analysis of regulations, environmental impacts and bioremediation systems. Ocean & Coastal Management 52, 243–249. https://doi.org/10.1016/j.ocecoaman.2009.03.002.
Bustamante, R.O., Castor, C., 1998. The decline of an endangered temperate ecosystem: the ruil (Nothofagus alessandrii) forest in central Chile. Biodiversity and Conservation 7, 1607–1626. https://doi.org/10.1023/A:1008856912888.
Bustamante, R.O., Simonetti, J.A., 2005. Is Pinus radiata invading the native vegetation in central Chile? Demographic responses in a fragmented forest. Biol Invasions 7, 243–249. https://doi.org/10.1007/s10530-004-0740-5.
Bustillos Ardaya, A., Evers, M., Ribbe, L., 2017. What influences disaster risk perception? Intervention measures, flood and landslide risk perception of the population living in flood risk areas in Rio de Janeiro state, Brazil. International Journal of Disaster Risk Reduction 25, 227–237. https://doi.org/10.1016/j.ijdrr.2017.09.006.

Caballé, G., Fernández, M.E., Gyenge, J., Lantschner, V., Rusch, V., Leourneau, F., Borrelli, L., 2016. Silvopastoral Systems Based on Natural Grassland and Ponderosa Pine in Northwestern Patagonia, Argentina, in: Peri, P.L., Dube, F., Varella, A. (Eds.), Silvopastoral Systems in Southern South America, 1st ed. Springer International Publishing; Springer e-books; Imprint; Springer, Cham, pp. 89–116.

Cairns, R., Stirling, A., 2014. 'Maintaining planetary systems' or 'concentrating global power?' High stakes in contending framings of climate geoengineering. Global Environmental Change 28, 25–38. https://doi.org/10.1016/j.gloenvcha.2014.04.005.

Camus, P., 2003. Federico Albert: Artífice de la Gestión de los Bosques de Chile. RGNG, 55–63.

Candan, F., Broquen, P., 2009. Aggregate stability and related properties in NW Patagonian Andisols. Geoderma 154, 42–47. https://doi.org/10.1016/j.geoderma.2009.09.010.

Carranza, D.M., Varas-Belemmi, K., Veer, D. de, Iglesias-Müller, C., Coral-Santacruz, D., Méndez, F.A., Torres-Lagos, E., Squeo, F.A., Gaymer, C.F., 2020. Socio-environmental conflicts: An underestimated threat to biodiversity conservation in Chile. Environmental Science & Policy 110, 46–59. https://doi.org/10.1016/j.envsci.2020.04.006.

Carrasco, L.R., Chan, J., McGrath, F.L., Le Nghiem, T.P., 2017. Biodiversity conservation in a telecoupled world. E&S 22. https://doi.org/10.5751/ES-09448-220324.

Carretier, S., Tolorza, V., Regard, V., Aguilar, G., Bermúdez, M.A., Martinod, J., Guyot, J.-L., Hérail, G., Riquelme, R., 2018. Review of erosion dynamics along the major N-S climatic gradient in Chile and perspectives. Geomorphology 300, 45–68. https://doi.org/10.1016/j.geomorph.2017.10.016.

Carruthers, D., 2001. Environmental politics in Chile: Legacies of dictatorship and democracy. Third World Quarterly 22, 343–358. https://doi.org/10.1080/01436590120061642.

Carruthers, D., Rodriguez, P., 2009a. Mapuche Protest, Environmental Conflict and Social Movement Linkage in Chile. Third World Quarterly 30, 743–760. https://doi.org/10.1080/01436590902867193.

Carruthers, D., Rodriguez, P., 2009b. Mapuche Protest, Environmental Conflict and Social Movement Linkage in Chile. Third World Quarterly 30, 743–760.

Casanova, M., Salazar, O., Seguel, O., Luzio, W., 2013. The Soils of Chile. Springer Netherlands.

Castellaro, G., Morales, L., Rodrigo, P., Fuentes, G., 2016. Carga ganaderay capacidad de carga de los pastizales naturales de la Patagonia chilena: estimación a nivel comunal. Agr Sur 44, 93–105.

Cawsey, E.M., Freudenberger, D., 2008. Assessing the biodiversity benefits of plantations: The Plantation Biodiversity Benefits Score. Ecol Manage Restor 9, 42–52. https://doi.org/10.1111/j.1442-8903.2008.00386.x.

Cembrano, J., Lara, L., 2009. The link between volcanism and tectonics in the southern volcanic zone of the Chilean Andes: A review. Tectonophysics 471, 96–113. https://doi.org/10.1016/j.tecto.2009.02.038.

2002. Censo 2002: Sintesis de Resultados: Report of Chilean Statistical Authorities, Santiago de Chile, Chile.

Cernev, T., 2022. Global sustainability targets: Planetary boundary, global catastrophic risk, and disaster risk reduction considerations. Progress in Disaster Science 16, 100264. https://doi.org/10.1016/j.pdisas.2022.100264.

Challies, E., Newig, J., Lenschow, A., 2019. Governance for Sustainability in Telecoupled Systems, in: Telecoupling. Palgrave Macmillan, Cham, pp. 177–197.

Chan, K.M.A., Boyd, D.R., Gould, R.K., Jetzkowitz, J., Liu, J., Muraca, B., Naidoo, R., Olmsted, P., Satterfield, T., Selomane, O., Singh, G.G., Sumaila, R., Ngo, H.T., Boedhihartono, A.K., Agard, J., Aguiar, A.P.D. de, Armenteras, D., Balint, L., Barrington-Leigh, C., Cheung, W.W.L., Díaz, S., Driscoll, J., Esler, K., Eyster, H., Gregr, E.J., Hashimoto, S., Hernández Pedraza, G.C., Hickler, T., Kok, M., Lazarova, T., Mohamed, A.A.A., Murray-Hudson, M., O'Farrell, P., Palomo, I., Saysel, A.K., Seppelt, R., Settele, J., Strassburg, B., Xue, D., Brondízio, E.S., 2020. Levers and leverage points for pathways to sustainability. People and Nature 2, 693–717. https://doi.org/10.1002/pan3.10124.

Chen, C.W., Gilmore, M., 2015. Biocultural Rights: A New Paradigm for Protecting Natural and Cultural Resources of Indigenous Communities. iipj 6. https://doi.org/10.18584/iipj.2015.6.3.3.

Chey, V.K., Holloway, J.D., Speight, M.R., 1997. Diversity of moths in forest plantations and natural forests in Sabah. Bull. Entomol. Res. 87, 371–385. https://doi.org/10.1017/S00 0748530003738X.

Chillo, V., Ladio, A.H., Salinas Sanhueza, J., Soler, R., Arpigiani, D.F., Rezzano, C.A., Cardozo, A.G., Peri, P.L., Amoroso, M.M., 2021. Silvopastoral Systems in Northern Argentine-Chilean Andean Patagonia: Ecosystem Services Provision in a Complex Territory, in: Peri, P.L., Pastur, G.M., Nahuelhual, L. (Eds.), Ecosystem Services in Patagonia: A Multi-Criteria Approach for an Integrated Assessment. Springer, Cham.

Chiodelli, F., Moroni, S., 2015. Corruption in land-use issues: a crucial challenge for planning theory and practice. Town Planning Review 86, 437–455. https://doi.org/10.3828/tpr.2015.27.

Christmann, G., Balgar, K., Mahlkow, N., 2014. Local Constructions of Vulnerability and Resilience in the Context of Climate Change. A Comparison of Lübeck and Rostock. Social Sciences 3, 142–159. https://doi.org/10.3390/socsci3010142.

Cid-Aguayo, B.E., Krstulovic-Matus, J.E., Carrasco Henríquez, N., Mella-Moraga, V., Oñate Vargas, D., 2022. Callampas of disaster: negotiations and struggles for the commons under forestry hegemony in Chile. Community Development Journal, bsac030. https://doi.org/10.1093/cdj/bsac030.

Clapp, R.A., 1995a. Creating Competitive Advantage: Forest Policy as Industrial Policy in Chile. Economic Geography 71, 273. https://doi.org/10.2307/144312.

Clapp, R.A., 1995b. The Unnatural History of the Monterey Pine. Geographical Review 85, 1. https://doi.org/10.2307/215551.

Clapp, R.A., 2001. Tree Farming and Forest Conservation in Chile: Do Replacement Forests Leave Any Originals Behind? Society & Natural Resources 14, 341–356. https://doi.org/10.1080/08941920119176.

2021. climate-data.org: ECMWF. https://de.climate-data.org/suedamerika/chile/region-aysen-del-general-carlos-ibanez-del-campo/coyhaique-2055/ (accessed 16 March 2021).

CONAF, 2015. Politica forestal 2015–2035, 76 pp.

Conedera, M., Torriani, D., Neff, C., Ricotta, C., Bajocco, S., Pezzatti, G.B., 2011. Using Monte Carlo simulations to estimate relative fire ignition danger in a low-to-medium fire-prone region. Forest Ecology and Management 261, 2179–2187. https://doi.org/10.1016/j.foreco.2010.08.013.

Conget, C., Núñez-Ávila, M. Hacia un nuevo Modelo Forestal en Chile.

Corbin, J., 2011. Grounded Theory, in: Bohnsack, R., Marotzki, W., Meuser, M. (Eds.), Hauptbegriffe qualitativer Sozialforschung, 3rd ed. Verlag Barbara Budrich, Opladen, Farmington Hills, MI, pp. 70–75.

Corenblit, D., Baas, A.C., Bornette, G., Darrozes, J., Delmotte, S., Francis, R.A., Gurnell, A.M., Julien, F., Naiman, R.J., Steiger, J., 2011. Feedbacks between geomorphology and biota controlling Earth surface processes and landforms: A review of foundation concepts and current understandings. Earth-Science Reviews 106, 307–331. https://doi.org/10.1016/j.earscirev.2011.03.002.

Corley, J.C., Villacide, J.M., Vesterinen, M., 2012. Can early thinning and pruning lessen the impact of pine plantations on beetle and ant diversity in the Patagonian steppe? Southern Forests: a Journal of Forest Science 74, 195–202. https://doi.org/10.2989/20702620.2012.722837.

Costantini, E.A.C., Branquinho, C., Nunes, A., Schwilch, G., Stavi, I., Valdecantos, A., Zucca, C., 2016. Soil indicators to assess the effectiveness of restoration strategies in dryland ecosystems. Solid Earth 7, 397–414. https://doi.org/10.5194/se-7-397-2016.

Cox, M., Arnold, G., Villamayor-Tomás, S., 2010. A Review of Design Principles for Community-based Natural Resource Management. E&S 15.

Cretney, R., 2019. "An opportunity to hope and dream": Disaster Politics and the Emergence of Possibility through Community-Led Recovery. Antipode 51, 497–516. https://doi.org/10.1111/anti.12431.

Cumming, G.S., 2011a. Spatial resilience in social-ecological systems, 1st ed. Springer, Dordrecht, Heidelberg, 254 pp.

Cumming, G.S., 2011b. Spatial resilience: integrating landscape ecology, resilience, and sustainability. Landscape Ecol 26, 899–909. https://doi.org/10.1007/s10980-011-9623-1.

Cutter, S.L., 2006. Hazards, vulnerability and environmental justice. Earthscan, London, Sterling, Va.

Cutter, S.L., Ash, K.D., Emrich, C.T., 2014. The geographies of community disaster resilience. Global Environmental Change 29, 65–77. https://doi.org/10.1016/j.gloenvcha.2014.08.005.

Czarniawska-Joerges, B., 2009. A theory of organising. Edward Elgar, Cheltenham, 153 pp.

Dadashpoor, H., Heydari, M., 2022. Schools of urban and regional planning evaluation: A genealogical perspective. IRSPSD International 10, 302–320. https://doi.org/10.14246/irspsd.10.3_302.

Dannenberg, A., Diekert, F., Händel, P., 2022. The effects of social information and luck on risk behavior of small-scale fishers at Lake Victoria. Journal of Economic Psychology 90, 102493. https://doi.org/10.1016/j.joep.2022.102493.

Datola, G., 2023. Implementing urban resilience in urban planning: A comprehensive framework for urban resilience evaluation. Sustainable Cities and Society 98, 104821. https://doi.org/10.1016/j.scs.2023.104821.

Davoudi, S., 2003. EUROPEAN BRIEFING: Polycentricity in European spatial planning: from an analytical tool to a normative agenda. European Planning Studies 11, 979–999. https://doi.org/10.1080/0965431032000146169.

Dawley, S., Pike, A., Tomaney, J., 2010. Towards the Resilient Region? Local Economy 25, 650–667. https://doi.org/10.1080/02690942.2010.533424.

References

Dec, D., Dörner, J., Balocchi, O., 2011. Temporal and spatial variability of structure dependent properties of a volcanic ash soil under pasture in southern Chile. Chilean J. Agric. Res. 71, 293–303. https://doi.org/10.4067/S0718-58392011000200015.

Dec, D., Dörner, J., Balocchi, O., López, I., 2012. Temporal dynamics of hydraulic and mechanical properties of an Andosol under grazing. Soil and Tillage Research 125, 44–51. https://doi.org/10.1016/j.still.2012.05.018.

Del Aigo, J.C., Skewes, J.C., Bañales-Seguel, C., Riquelme Maulén, W., Molares, S., Morales, D., Ibarra, M.I., Guerra, D., 2022. Waterscapes In Wallmapu: Lessons From Mapuche Perspectives. Geographical Review 112, 622–640. https://doi.org/10.1080/001 67428.2020.1800410.

Delamaza, G., Maillet, A., Martínez Neira, C., 2017. Socio-Territorial Conflicts in Chile: Configuration and Politicization (2005–2014). ERLACS 0, 23. https://doi.org/10.18352/erlacs.10173.

DellaSala, D.A., 2020. "Real" vs. "Fake" Forests: Why Tree Plantations Are Not Forests, in: Goldstein, M.I., DellaSala, D.A. (Eds.), Encyclopedia of the world's biomes. Elsevier, pp. 47–55.

Devisscher, T., Spies, J., Griess, V.C., 2021. Time for change: Learning from community forests to enhance the resilience of multi-value forestry in British Columbia, Canada. Land Use Policy 103, 105317. https://doi.org/10.1016/j.landusepol.2021.105317.

Dexter, A.R., 1988. Advances in characterization of soil structure. Soil and Tillage Research 11, 199–238. https://doi.org/10.1016/0167-1987(88)90002-5.

Di Giminiani, P., 2016a. How to Manage a Forest: Environmental Governance in Neoliberal Chile. Anthropological Quarterly 89, 723–751.

Di Giminiani, P., 2016b. How to Manage a Forest: Environmental Governance in Neoliberal Chile. Anthropological Quarterly 89, 723–751.

Di Giminiani, P., Fonck, M., 2018. Emerging landscapes of private conservation: Enclosure and mediation in southern Chilean protected areas. Geoforum 97, 305–314. https://doi.org/10.1016/j.geoforum.2018.09.018.

Di Giminiani, P., González Gálvez, M., 2018. Who Owns the Water? The Relation as Unfinished Objectivation in the Mapuche Lived World. Anthropological Forum 28, 199–216. https://doi.org/10.1080/00664677.2018.1495060.

Diehl, A., Abdul-Rahman, A., Bach, B., El-Assady, M., Kraus, M., Laramee, R.S., Keim, D.A., Chen, M., 2022. Characterizing Grounded Theory Approaches in Visualization, 13 pp. http://arxiv.org/pdf/2203.01777v2.

Dintwa, K.F., Letamo, G., Navaneetham, K., 2022. Vulnerability perception, quality of life, and indigenous knowledge: A qualitative study of the population of Ngamiland West District, Botswana. International Journal of Disaster Risk Reduction 70, 102788. https://doi.org/10.1016/j.ijdrr.2022.102788.

División de Planificación y Desarrollo Regional, 2013. Plan Regional de Ordenamiento Territorial de Aysén. https://www.goreaysen.cl/controls/neochannels/neo_ch95/appins tances/media204/Plan_Regional_de_Ordenamiento_Territorial_de_Aysen.pdf (accessed 6 November 2023).

División de Planificación y Desarrollo Regional, 2015. Plan Regional de Ordenamiento Territorial: Zonificación Regional. GOBIERNO REGIONAL DE AYSÉN.

Dlamini, C., Montouroy, Y., 2017. Governing Sustainable Forest Management Issues in Polycentric Governance. Environmental Law Review 19, 6–29. https://doi.org/10.1177/146 1452917691863.

Dombrowsky, W.R., 2008. Zur Entstehung der soziologischen Katastrophenforschung - eine wissenshistorische und -soziologische Reflexion, in: Felgentreff, C., Glade, T. (Eds.), Naturrisiken und Sozialkatastrophen. Spektrum Akad. Verl., Berlin, pp. 63–76.

Dominicis, S. de, Fornara, F., Ganucci Cancellieri, U., Twigger-Ross, C., Bonaiuto, M., 2015. We are at risk, and so what? Place attachment, environmental risk perceptions and preventive coping behaviours. Journal of Environmental Psychology 43, 66–78. https://doi.org/10.1016/j.jenvp.2015.05.010.

Dörner, J., Dec, D., Feest, E., Vásquez, N., Díaz, M., 2012. Dynamics of soil structure and pore functions of a volcanic ash soil under tillage. Soil and Tillage Research 125, 52–60. https://doi.org/10.1016/j.still.2012.05.019.

Dörner, J., Dec, D., Peng, X., Horn, R., 2009. Change of shrinkage behavior of an Andisol in southern Chile: Effects of land use and wetting/drying cycles. Soil and Tillage Research 106, 45–53. https://doi.org/10.1016/j.still.2009.09.013.

Douglas, M., Wildavsky, A.B., 1983. Risk and culture: An essay on the selection of technological and environmental dangers. Univ. of Calif. Pr, Berkeley, Calif. [u.a.], IX, 221.

Draucker, C.B., Martsolf, D.S., Ross, R., Rusk, T.B., 2007. Theoretical sampling and category development in grounded theory. Qualitative health research 17, 1137–1148. https://doi.org/10.1177/1049732307308450.

Dressler and BÜSCHER, B., SCHOON, M., BROCKINGTON, D.A., HAYES, T., KULL, C.A., McCarthy, J., SHRESTHA, K., , 2010.Dressler, W., BÜSCHER, B., SCHOON, M., BROCKINGTON, D.A., HAYES, T., KULL, C.A., McCarthy, J., SHRESTHA, K., 2010. From hope to crisis and back again? A critical history of the global CBNRM narrative. Envir. Conserv. 37, 5–15. https://doi.org/10.1017/S0376892910000044.

Dube, F., Zagal, E., Stolpe, N., Espinosa, M., 2009. The influence of land-use change on the organic carbon distribution and microbial respiration in a volcanic soil of the Chilean Patagonia. Forest Ecology and Management 257, 1695–1704. https://doi.org/10.1016/j.foreco.2009.01.044.

Echeverria, C., Coomes, D., Salas, J., Rey-Benayas, J.M., Lara, A., Newton, A., 2006. Rapid deforestation and fragmentation of Chilean Temperate Forests. Biological Conservation 130, 481–494. https://doi.org/10.1016/j.biocon.2006.01.017.

Eckelmann, H.W., Sponagel, R.:.H., Grottenthaler, W., Hartmann, K.-J., Hartwich, R., Janetzko, P., Joisten, H., Kühn, D., Sabel, K.-J., Traidl, R., Boden, H.:.A.-A. (Eds.), 2006. Bodenkundliche Kartieranleitung. KA5. Schweizerbart Science Publishers, Stuttgart, Germany.

Eichel, J., Corenblit, D., Dikau, R., 2016. Conditions for feedbacks between geomorphic and vegetation dynamics on lateral moraine slopes: a biogeomorphic feedback window. Earth Surf. Process. Landforms 41, 406–419. https://doi.org/10.1002/esp.3859.

Eichel, J., Draebing, D., Klingbeil, L., Wieland, M., Eling, C., Schmidtlein, S., Kuhlmann, H., Dikau, R., 2017. Solifluction meets vegetation: the role of biogeomorphic feedbacks for turf-banked solifluction lobe development. Earth Surf. Process. Landforms 42, 1623–1635. https://doi.org/10.1002/esp.4102.

Eichel, J., Draebing, D., Meyer, N., 2018. From active to stable: Paraglacial transition of Alpine lateral moraine slopes. Land Degrad Dev 29, 4158–4172. https://doi.org/10.1002/ldr.3140.

Eichel, J., Krautblatter, M., Schmidtlein, S., Dikau, R., 2013. Biogeomorphic interactions in the Turtmann glacier forefield, Switzerland. Geomorphology 201, 98–110. https://doi.org/10.1016/j.geomorph.2013.06.012.

Eijkelkamp. Pocket Vane Tester. https://en.eijkelkamp.com/products/field-measurement-equipment/pocket-vane-tester.html (accessed 8 January 2020).

El-Hani, C.N., Poliseli, L., Ludwig, D., 2022. Beyond the divide between indigenous and academic knowledge: Causal and mechanistic explanations in a Brazilian fishing community. Studies in History and Philosophy of Science 91, 296–306. https://doi.org/10.1016/j.shpsa.2021.11.001.

Escalona Ulloa, M., Barton, J.R., 2020. A "landscapes of power" framework for historical political ecology: The production of cultural hegemony in Araucanía-Wallmapu. Area 52, 445–454. https://doi.org/10.1111/area.12591#.

Estades, C.F., Grez, A.A., Simonetti, J.A., 2012. Biodiversity in Monterrey Pine Plantations, in: Simonetti, J.A., Audrey, A., Estades, C.F. (Eds.), Biodiversity conservation in agroforestry landscapes: challenges and opportunities. Editorial Universitaria.

Estades, C.F., Temple, S.A., 1999. Deciduous-Forest Bird Communities in a Fragmented Landscape Dominated by Exotic Pine Plantations. Ecological Applications 9, 573. https://doi.org/10.2307/2641145.

Fahrig, L., 2003. Effects of Habitat Fragmentation on Biodiversity. Annu. Rev. Ecol. Evol. Syst. 34, 487–515. https://doi.org/10.1146/annurev.ecolsys.34.011802.132419.

Fajardo, A., Gundale, M.J., 2015. Combined effects of anthropogenic fires and land-use change on soil properties and processes in Patagonia, Chile. Forest Ecology and Management 357, 60–67. https://doi.org/10.1016/j.foreco.2015.08.012.

Fajardo, A., Llancabure, J.C., Moreno, P.C., 2022. Assessing forest degradation using multivariate and machine-learning methods in the Patagonian temperate rain forest. Ecological applications : a publication of the Ecological Society of America 32, e2495. https://doi.org/10.1002/eap.2495.

Faria, D., Mariano-Neto, E., Martini, A.M.Z., Ortiz, J.V., Montingelli, R., Rosso, S., Paciencia, M.L.B., Baumgarten, J., 2009. Forest structure in a mosaic of rainforest sites: The effect of fragmentation and recovery after clear cut. Forest Ecology and Management 257, 2226–2234. https://doi.org/10.1016/j.foreco.2009.02.032.

Farjon, A., 2010. A handbook of the world's conifers. Koninklijke Brill NV, Leiden, Boston.

Feitelson, E., Felsenstein, D., Razin, E., Stern, E., 2017. Assessing land use plan implementation: Bridging the performance-conformance divide. Land Use Policy 61, 251–264. https://doi.org/10.1016/j.landusepol.2016.11.017.

Felgentreff, C., 2003. Raumplanung in der Naturgefahren- und Risikoforschung. Universitätsverlag, Potsdam, 89 pp.

Felgentreff, C., Glade, T. (Eds.), 2008. Naturrisiken und Sozialkatastrophen. Spektrum Akad. Verl., Berlin, 454 pp.

Ferns, P., Cowie, R., Weir, A., 1992. Managing conifer plantations for the benefit of granivorous birds and mammals. Aspects Appl. Biol. 29, 83–88.

Festinger, L., 2001. A theory of cognitive dissonance, 1962nd ed. Stanford Univ. Press, Stanford, 291 pp.

Feurdean, A., Vannière, B., Finsinger, W., Warren, D., Connor, S.C., Forrest, M., Liakka, J., Panait, A., Werner, C., Andrič, M., Bobek, P., Carter, V.A., Davis, B., Diaconu, A.-C., Dietze, E., Feeser, I., Florescu, G., Gałka, M., Giesecke, T., Jahns, S., Jamrichová, E.,

Kajukało, K., Kaplan, J., Karpińska-Kołaczek, M., Kołaczek, P., Kuneš, P., Kupriyanov, D., Lamentowicz, M., Lemmen, C., Magyari, E.K., Marcisz, K., Marinova, E., Niamir, A., Novenko, E., Obremska, M., Pędziszewska, A., Pfeiffer, M., Poska, A., Rösch, M., Słowiński, M., Stančikaitė, M., Szal, M., Święta-Musznicka, J., Tanţău, I., Theuerkauf, M., Tonkov, S., Valkó, O., Vassiljev, J., Veski, S., Vincze, I., Wacnik, A., Wiethold, J., Hickler, T., 2020. Fire hazard modulation by long-term dynamics in land cover and dominant forest type in eastern and central Europe. Biogeosciences 17, 1213–1230. https://doi.org/10.5194/bg-17-1213-2020.

Fischer, J., Lindenmayer, D.B., 2007. Landscape modification and habitat fragmentation: a synthesis. Global Ecol Biogeography 16, 265–280. https://doi.org/10.1111/j.1466-8238.2007.00287.x.

Fischer, J., Riechers, M., 2019. A leverage points perspective on sustainability. People Nat 1, 115–120. https://doi.org/10.1002/pan3.13.

Foley, J.A., Defries, R., Asner, G.P., Barford, C., Bonan, G., Carpenter, S.R., Chapin, F.S., Coe, M.T., Daily, G.C., Gibbs, H.K., Helkowski, J.H., Holloway, T., Howard, E.A., Kucharik, C.J., Monfreda, C., Patz, J.A., Prentice, I.C., Ramankutty, N., Snyder, P.K., 2005. Global consequences of land use. Science (New York, N.Y.) 309, 570–574. https://doi.org/10.1126/science.1111772.

Fomegas, David, Y., Bayos, Norma, W., Atayoc, J.P., 2004. Phytodiversity Under Pine (Pinus insularis) Forest Community In Tadian, Mountain Province 1. Regional Sectoral/Commodity Review.

França, L.C.d.J., Júnior, F.W.A., Jarochinski e Silva, C.S., Monti, C.A.U., Ferreira, T.C., Santana, C.J.d.O., Gomide, L.R., , 2022.França, L.C.d.J., Júnior, F.W.A., Jarochinski e Silva, C.S., Monti, C.A.U., Ferreira, T.C., Santana, C.J.d.O., Gomide, L.R., 2022. Forest landscape planning and management: A state-of-the-art review. Trees, Forests and People 8, 100275. https://doi.org/10.1016/j.tfp.2022.100275.

Franchi-Arzola, I., Martin-Vide, J., Henríquez, C., 2018. Sustainability Assessment in Development Planning in Sub-National Territories: Regional Development Strategies in Chile. Sustainability 10, 1398. https://doi.org/10.3390/su10051398.

Freedman, B., Zelazny, V., Beaudette, D., Fleming, T., Johnson, G., Flemming, S., Gerrow, J.S., Forbes, G., Woodley, S., 1996. Biodiversity implications of changes in the quantity of dead organic matter in managed forests. Environ. Rev. 4, 238–265. https://doi.org/10.1139/a96-013.

Friend, G.R., 1982. Bird Populations in Exotic Pine Plantations and Indigenous Eucalypt Forests in Gippsland, Victoria. Emu - Austral Ornithology 82, 80–91. https://doi.org/10.1071/MU9820080.

Fuentes-Ramírez, A., Pauchard, A., Cavieres, L.A., García, R.A., 2011. Survival and growth of Acacia dealbata vs. native trees across an invasion front in south-central Chile. Forest Ecology and Management 261, 1003–1009. https://doi.org/10.1016/j.foreco.2010.12.018.

Fukuda-Parr, S., Muchhala, B., 2020. The Southern origins of sustainable development goals: Ideas, actors, aspirations. World Development 126, 104706. https://doi.org/10.1016/j.worlddev.2019.104706.

Funk, K., 2012. "Today There Are No Indigenous People" in Chile?: Connecting the Mapuche Struggle to Anti-Neoliberal Mobilizations in South America. Journal of Politics in Latin America 4, 125–140. https://doi.org/10.1177/1866802X1200400205.

Fürst, D., Scholles, F., 2008. Handbuch Theorien und Methoden der Raum- und Umweltplanung, 3rd ed. Rohn, Dortmund, 656 pp.

Gaillard, J., 2019a. Petition Power, Prestige & Forgotten Values: A Disaster Studies Manifesto. https://www.ipetitions.com/petition/power-prestige-forgotten-values-a-disaster (accessed 11 October 2022).

Gaillard, J., 2019b. Petition Power, Prestige & Forgotten Values: A Disaster Studies Manifesto. https://www.ipetitions.com/petition/power-prestige-forgotten-values-a-disaster (accessed 11 October 2022).

Galaz, V., Biermann, F., Crona, B., Loorbach, D., Folke, C., Olsson, P., Nilsson, M., Allouche, J., Persson, Å., Reischl, G., 2012. 'Planetary boundaries'—exploring the challenges for global environmental governance. Current Opinion in Environmental Sustainability 4, 80–87. https://doi.org/10.1016/j.cosust.2012.01.006.

García, R.A., Franzese, J., Policelli, N., Sasal, Y., ZENNI, R.D., Nuñez, M.A., Taylor, K., Pauchard, A., 2018. Non-native Pines Are Homogenizing the Ecosystems of South America, in: From Biocultural Homogenization to Biocultural Conservation. Springer, Cham, pp. 245–263.

García, R.A., Pauchard, A., Escudero, A., 2014. French broom (Teline monspessulana) invasion in south-central Chile depends on factors operating at different spatial scales. Biol Invasions 16, 113–124. https://doi.org/10.1007/s10530-013-0507-y.

Garvey, R., Silva Carrasco, E., Roa Solís, C., Charó Bortolaso, C., 2023. Prehistoric Human Occupation of Southern Andean Forests: Evidence from Alero Largo, Aysén, Chilean Patagonia. Latin Am. antiq. 34, 366–384. https://doi.org/10.1017/laq.2022.18.

Gauthier, S., Bernier, P., Kuuluvainen, T., Shvidenko, A.Z., Schepaschenko, D.G., 2015. Boreal forest health and global change. Science (New York, N.Y.) 349, 819–822. https://doi.org/10.1126/science.aaa9092.

Geiger, R., Pohl, W., 1961. Eine neue Wandkarte der Klimagebiete der Erde. Erdkunde 8, 58–60.

Genet, M., Kokutse, N., Stokes, A., Fourcaud, T., Cai, X., Ji, J., Mickovski, S., 2008. Root reinforcement in plantations of Cryptomeria japonica D. Don: effect of tree age and stand structure on slope stability. Forest Ecology and Management 256, 1517–1526. https://doi.org/10.1016/j.foreco.2008.05.050.

Genske, D.D., 2006. Ingenieurgeologie: Grundlagen und Anwendung. Springer-Verlag Berlin Heidelberg, Berlin, Heidelberg.

Gerber, J.-F., 2011. Conflicts over industrial tree plantations in the South: Who, how and why? Global Environmental Change 21, 165–176. https://doi.org/10.1016/j.gloenvcha.2010.09.005.

Gerke, H.H., Vogel, H.-J., Weber, T.K., van der Meij, W.M., Scholten, T., 2022. 3–4D soil model as challenge for future soil research: Quantitative soil modeling based on the solid phase. J. Plant Nutr. Soil Sci. 185, 720–744. https://doi.org/10.1002/jpln.202200239.

Ghazoul, J., 2013. Deforestation and Land Clearing, in: Levin, S.A. (Ed.), Encyclopedia of biodiversity, 2nd ed. Academic Press, Amsterdam, pp. 447–456.

Ghosh, B., Ramos-Mejía, M., Machado, R.C., Yuana, S.L., Schiller, K., 2021. Decolonising transitions in the Global South: Towards more epistemic diversity in transitions research. Environmental Innovation and Societal Transitions 41, 106–109. https://doi.org/10.1016/j.eist.2021.10.029.

Gill, A.M., Williams, J.E., 1996. Fire regimes and biodiversity: the effects of fragmentation of southeastern Australian eucalypt forests by urbanisation, agriculture and pine plantations. Forest Ecology and Management 85, 261–278. https://doi.org/10.1016/S0378-1127(96)03763-2.

Gill, J.C., Malamud, B.D., 2017. Anthropogenic processes, natural hazards, and interactions in a multi-hazard framework. Earth-Science Reviews 166, 246–269. https://doi.org/10.1016/j.earscirev.2017.01.002.

Glæsner, N., Helming, K., Vries, W. de, 2014. Do Current European Policies Prevent Soil Threats and Support Soil Functions? Sustainability 6, 9538–9563. https://doi.org/10.3390/su6129538.

Glasser, N.F., Jansson, K.N., Harrison, S., Kleman, J., 2008. The glacial geomorphology and Pleistocene history of South America between 38°S and 56°S. Quaternary Science Reviews 27, 365–390. https://doi.org/10.1016/j.quascirev.2007.11.011.

Gobierno Regional de Aysén, 2013. Actualización del Plan Regional de Ordenamiento Territorial de Aysén: Memoria Explicativa. Unidad de Planificación Territorial y Borde Costero, Coyhaique. http://www.goreaysen.cl/controls/neochannels/neo_ch95/appinstances/media204/Plan_Regional_de_Ordenamiento_Territorial_de_Aysen.pdf.

Gobierno Regional de Aysén, 2019. Plan de Prevención y Protección ante Situaciones de Emergencia y Desastres Naturales y Antrópicos, Coyhaique, Chile, 157 pp. https://www.goreaysen.cl/controls/neochannels/neo_ch95/appinstances/media204/Plan_de_Prevencion_y_Proteccion_ante_Situaciones_de_Desastre_INFORME_vf.pdf (accessed 6 November 2023).

Goetz, A., Hussein, H., Thiel, A., 2023. Polycentric governance and agroecological practices in the MENA region: insights from Lebanon, Morocco and Tunisia. International Journal of Water Resources Development, 1–16. https://doi.org/10.1080/07900627.2023.2260902.

Gómez, P., Hahn, S., San Martín, J., 2009. Estructura Y Composición Florística De Un Matorral Bajo Plantaciones De Pinus Radiata D.Don En Chile Central. Gayana Bot. 66. https://doi.org/10.4067/S0717-66432009000200010.

González, D.P., 2022. Disaster Risk Governance as Assemblage: The Chilean Framework of the 1985 San Antonio Earthquake. Int J Disaster Risk Sci 13, 878–889. https://doi.org/10.1007/s13753-022-00453-y.

Greenway, D.R., 1987. Vegetation and slope stability, in: Anderson M.G., Richards K. S. (Ed.), Slope stability, Geotechnical Engineering and Geomorphology. John Wiley & Sons, Ltd, pp. 198–230.

Greiving, S., 2002. Räumliche Planung und Risiko. Gerling Akad. Verl., München, 320 pp.

Greiving, S., 2011. Methodik zur Festlegung raum- und raumplanungsrelevanter Risiken, in: Pohl, J., Zehetmair, S. (Eds.), Risikomanagement als Handlungsfeld in der Raumplanung. Hannover, pp. 22–30.

Greiving, S., Fleischhauer, M., Wanczura, S., 2006. Management of natural hazards in Europe: The role of spatial planning in selected EU member states. Journal of Environmental Planning and Management 49, 739–757. https://doi.org/10.1080/09640560600850044.

Greiving, S., Kruse, P., Othmer, F., Fleischhauer, M., Fuchs, M., 2023. Implementation of Risk-Based Approaches in Urban Land Use Planning—The Example of the City of Erftstadt, Germany. Sustainability 15, 15340. https://doi.org/10.3390/su152115340.

Greiving, S., Pohl, J., 2011. Anwendungsbeispiel: Hangrutschungen (in der Schwäbischen Alb), in: Pohl, J., Zehetmair, S. (Eds.), Risikomanagement als Handlungsfeld in der Raumplanung. Hannover, pp. 71–75.
Grime, J.P., 2002. Plant strategies, vegetation processes and properties, 2nd ed. John Wiley & Sons, Wiley.
Guerrero, P.C., Bustamante, R.O., 2007. Can native tree species regenerate in Pinus radiata plantations in Chile? Forest Ecology and Management 253, 97–102. https://doi.org/10.1016/j.foreco.2007.07.006.
Gut, B., 2008. Trees in Patagonia. Birkhäuser Verlag AG, Basel, Boston, Berlin.
Gutmann, A., 2023. ¿Pachamama como sujeto de derecho? Derechos de la Naturaleza y pensamiento indígena en Ecuador, in: Fischer-Lescano, A., Franco, A.V. (Eds.), La Naturaleza como sujeto de derechos: Un diálogo folosófico y jurídico entre alemania y ecuador. El Siglo, pp. 151–172.
Gwynne, R.N., 1996. Direct Foreign Investment and Non-traditional Export Growth in Chile: the Case of the Forestry Sector. Bull Latin American Research 15, 341–357. https://doi.org/10.1111/j.1470-9856.1996.tb00041.x.
Gyenge, J.E., Fernndez, M.E., Schlichter, T.M., 2003. Water relations of ponderosa pines in Patagonia Argentina: implications for local water resources and individual growth. Trees - Structure and Function 17, 417–423. https://doi.org/10.1007/s00468-003-0254-2.
Haberzettl, T., Wille, M., Fey, M., Janssen, S., Lücke, A., Mayr, C., Ohlendorf, C., Schäbitz, F., Schleser, G.H., Zolitschka, B., 2006. Environmental change and fire history of southern Patagonia (Argentina) during the last five centuries. Quaternary International 158, 72–82. https://doi.org/10.1016/j.quaint.2006.05.029.
Hamilton, R.T.V., Ramcilovic-Suominen, S., 2023. From hegemony-reinforcing to hegemony-transcending transformations: horizons of possibility and strategies of escape. Sustain Sci 18, 737–748. https://doi.org/10.1007/s11625-022-01257-1.
Hammer, C.C., Brainard, J., Innes, A., Hunter, P.R., 2019. (Re-) conceptualising vulnerability as a part of risk in global health emergency response: updating the pressure and release model for global health emergencies. Emerg Themes Epidemiol 16, 2. https://doi.org/10.1186/s12982-019-0084-3.
Haraway, D., 1988. Situated Knowledges: The Science Question in Feminism and the Privilege of Partial Perspective. Feminist Studies 14, 575. https://doi.org/10.2307/3178066.
Haraway, D.J., 1989. Primate visions, 1st ed. Routledge, New York, London, 486 pp.
Hartge, K.H., Horn, R., 2009. Die physikalische Untersuchung von Böden: [Praxis, Messmethoden, Auswertung], 4th ed. Schweizerbart, Stuttgart, 178 pp.
Hartmann, S., Pedoth, L., Dalla Torre, C., Schneiderbauer, S., 2021. Beyond the Expected—Residual Risk and Cases of Overload in the Context of Managing Alpine Natural Hazards. Int J Disaster Risk Sci 12, 205–219. https://doi.org/10.1007/s13753-020-00325-3.
Head, L., 2012. Conceptualising the human in cultural landscapes and resilience thinking, in: Plieninger, T., Bieling, C. (Eds.), Resilience and the cultural landscape: Understanding and managing change in human-shaped environments. Cambridge University Press, Cambridge, UK, New York, pp. 65–79.
Heaphy, M.J., Lowe, D.J., Palmer, D.J., Jones, H.S., Gielen, G.J., Oliver, G.R., Pearce, S.H., 2014. Assessing drivers of plantation forest productivity on eroded and non-eroded soils in hilly land, eastern North Island, New Zealand. N.Z. j. of For. Sci. 44. https://doi.org/10.1186/s40490-014-0024-5.

Heilmayr, R., Echeverría, C., Lambin, E.F., 2020. Impacts of Chilean forest subsidies on forest cover, carbon and biodiversity. Nat Sustain 3, 701–709. https://doi.org/10.1038/s41893-020-0547-0.

Heilmayr, R., Lambin, E.F., 2016. Impacts of nonstate, market-driven governance on Chilean forests. Proceedings of the National Academy of Sciences of the United States of America 113, 2910–2915. https://doi.org/10.1073/pnas.1600394113.

Heinberg, Richard and Miller, Asher (Ed.), 2023. Welcome to the Great Unraveling: Navigating the Polycrisis of Environmental and Social Breakdown. Post Carbon Institute, Corvallis, Oregon, 67 pp.

Heinze, T., 2016. Qualitative Sozialforschung: Einführung, Methodologie und Forschungspraxis, 2016th ed. Oldenbourg Wissenschaftsverlag, Berlin, Boston, 1 p.

Hernández Aguilar, B., Ruiz Rivera, N., 2016. The production of vulnerability to landslides: the risk habitus in two landslide-prone neighborhoods in Teziutlan, Mexico. Investigaciones Geográficas, Boletín del Instituto de Geografía 2016, 7–27. https://doi.org/10.14350/rig.50663.

Hernández-Moreno, Á., Soto, D.P., Miranda, A., Holz, A., Armenteras-Pascual, D., 2023. Forest landscape dynamics after intentional large-scale fires in western Patagonia reveal unusual temperate forest recovery trends. Landsc Ecol. https://doi.org/10.1007/s10980-023-01687-x.

Hetemäki, L., Seppälä, J., 2022. Planetary Boundaries and the Role of the Forest-Based Sector, in: Hetemäki, L., Kangas, J., Peltola, H. (Eds.), Forest bioeconomy and climate change. Springer, Cham, pp. 19–31.

Hildebrand-Vogel, R., Godoy, R., Vogel, A., 1990. Subantarctic-Andean Nothofagus pumilio forests. Vegetatio 89, 55–68. https://doi.org/10.1007/BF00134434.

Hiriart-Bertrand, L., Silva, J.A., Gelcich, S., 2020. Challenges and opportunities of implementing the marine and coastal areas for indigenous peoples policy in Chile. Ocean & Coastal Management 193, 105233. https://doi.org/10.1016/j.ocecoaman.2020.105233.

HMWEVW, 2023. Landesplanung Hessen. https://landesplanung.hessen.de/ (accessed 20 December 2023).

Holifield, R., Porter, M., Walker, G., Walker, G.P., Holifield, R.B. (Eds.), 2010. Spaces of environmental justice. Wiley-Blackwell, Chichester, West Sussex, U.K, Malden, MA, 263 pp.

Holmgren, M., Avilés, R., Sierralta, L., Segura, A.M., Fuentes, E.R., 2000. Why have European herbs so successfully invaded the Chilean matorral? Effects of herbivory, soil nutrients, and fire. Journal of Arid Environments 44, 197–211. https://doi.org/10.1006/jare.1999.0589.

Holz, A., Veblen, T.T., 2011. The amplifying effects of humans on fire regimes in temperate rainforests in western Patagonia. Palaeogeography, Palaeoclimatology, Palaeoecology 311, 82–92. https://doi.org/10.1016/j.palaeo.2011.08.008.

Hossain, M.S., Ramirez, J.A., Haisch, T., Speranza, C.I., Martius, O., Mayer, H., Keiler, M., 2020. A coupled human and landscape conceptual model of risk and resilience in Swiss Alpine communities. The Science of the total environment 730, 138322. https://doi.org/10.1016/j.scitotenv.2020.138322.

Huaiquimilla-Guerrero, K., Moscote-Guerra, J., Ramírez-Rivera, S., Duhalde-Correa, L.A., Peralta-Scholz, M.J., Silva, F.O., Santana, V.P., Barrera, A.S., Squeo, F.A., Gaymer, C.F., 2022. Dispossession and Governance: The Invisible Role of Indigenous Peoples in Protected Natural Areas in Chile.

Hübner, R., Böhm, C., Zehlius-Eckert, W., 2020. Rechtliche und politische Hemmnisse für die Agroforst-Wirtschaft: Lösungsvorschläge zu deren Überwindung, aktuelle Kompromisslösungen und besondere Fallstricke: Loseblatt #49, 32 pp. http://agroforst-info.de/.

Hudek, C., Stanchi, S., D'Amico, M., Freppaz, M., 2017. Quantifying the contribution of the root system of alpine vegetation in the soil aggregate stability of moraine. International Soil and Water Conservation Research 5, 36–42. https://doi.org/10.1016/j.iswcr.2017.02.001.

Hudson, R., 2010. Resilient regions in an uncertain world: wishful thinking or a practical reality? Cambridge Journal of Regions, Economy and Society 3, 11–25. https://doi.org/10.1093/cjres/rsp026.

Huertas Herrera, A., Promis, Á., Toro-Manríquez, M., Lencinas, M.V., Martínez Pastur, G., Río, M., 2022. Rehabilitation of Nothofagus pumilio forests in Chilean Patagonia: can fencing and planting season effectively protect against exotic European hare browsing? New Forests 53, 469–485. https://doi.org/10.1007/s11056-021-09867-w.

Hull, A., Alexander, E.R., Khakee, A., Woltjer, J. (Eds.), 2011. Evaluation for participation and sustainability in planning. Routledge, Abingdon, Oxon, 1 Online-Ressource.

Hurley, P.T., Walker, P.A., 2004. Whose Vision? Conspiracy Theory and Land-Use Planning in Nevada County, California. Environ Plan A 36, 1529–1547. https://doi.org/10.1068/a36186.

Ibáñez Santa María, A., 1973. La incorporación de Aisén a la vida nacional, 1902–1936. Pontificia Universidad Católica de Chile, Santiago, Chile.

Ibrahim, A.-S., Abubakari, M., Kepe, T., 2023. Land as Common Property: The Fit of Land Governance with Ostrom's Design Principles. Soc, 1–15. https://doi.org/10.1007/s12115-023-00913-1.

Iizuka, M., Roje, P., Vera, V., 2016. The Development of Salmon Aquaculture in Chile into an Internationally Competitive Industry: 1985–2007, in: Hosono, A., Iizuka, M., Katz, J. (Eds.), Chile's Salmon Industry: Policy Challenges in Managing Public Goods. Springer Japan, Tokyo, s.l., pp. 75–108.

Iizuka, M., Zanlungo, J.P., 2016. Environmental Collapse and Institutional Restructuring: The Sanitary Crisis in the Chilean Salmon Industry, in: Hosono, A., Iizuka, M., Katz, J. (Eds.), Chile's Salmon Industry: Policy Challenges in Managing Public Goods. Springer Japan, Tokyo, s.l., pp. 109–136.

Imaizumi, F., Sidle, R.C., Kamei, R., 2008. Effects of forest harvesting on the occurrence of landslides and debris flows in steep terrain of central Japan. Earth Surface Processes and Landforms 33, 827–840. https://doi.org/10.1002/esp.1574.

Inostroza, L., Zasada, I., König, H.J., 2016. Last of the wild revisited: assessing spatial patterns of human impact on landscapes in Southern Patagonia, Chile. Reg Environ Change 16, 2071–2085. https://doi.org/10.1007/s10113-016-0935-1.

IPCC (Ed.), 2022. Climate Change 2022: Impacts, Adaptation and Vulnerability: Contribution of Working Group II to the Sixth Assessment Report of the Intergovernmental Panel on Climate Change. Cambridge University Press, New York, USA, 3068 pp.

IPCC, 2023. Climate Change 2023: Synthesis Report. Contribution of Working Groups I, II and III to the Sixth Assessment Report of the Intergovernmental Panel on Climate Change [Core Writing Team, H. Lee and J. Romero (eds.)]., Geneva, Switzerland.

Javier E. Gyenge, María Elena Fernández, Verónica Rusch, Mauro M. Sarasola, Tomás M. Schlichter, 2010. Towards Sustainable Forestry Development in Patagonia: Truths and Myths of Environmental Impacts of Plantations with Fast-Growing Conifers. The Americas Journal of Plant Science and Biotechnology, 9–22.

Jeetze, P.J. von, Weindl, I., Johnson, J.A., Borrelli, P., Panagos, P., Molina Bacca, E.J., Karstens, K., Humpenöder, F., Dietrich, J.P., Minoli, S., Müller, C., Lotze-Campen, H., Popp, A., 2023. Projected landscape-scale repercussions of global action for climate and biodiversity protection. Nat Commun 14, 2515. https://doi.org/10.1038/s41467-023-380 43-1.

Johnson, J.A., Brown, M.E., Corong, E., Dietrich, J.P., C Henry, R., Jeetze, P.J. von, Leclère, D., Popp, A., Thakrar, S.K., Williams, D.R., 2023. The meso scale as a frontier in interdisciplinary modeling of sustainability from local to global scales. Environ. Res. Lett. 18, 25007. https://doi.org/10.1088/1748-9326/acb503.

Jullian, C., Nahuelhual, L., 2021. Land Size, Native Forests, and Ecosystem Service Inequalities in the Rural Chilean Patagonia, in: Peri, P.L., Pastur, G.M., Nahuelhual, L. (Eds.), Ecosystem Services in Patagonia: A Multi-Criteria Approach for an Integrated Assessment. Springer, Cham, pp. 379–396.

Kahneman, D., Tversky, A., 1974. Judgement under Uncertainty: Heuristics and Biases: Biases in judgments reveal some heuristics of thinking under uncertainty. Science, 1124–1131.

Kamarinas, I., Julian, J., Hughes, A., Owsley, B., Beurs, K. de, 2016. Nonlinear Changes in Land Cover and Sediment Runoff in a New Zealand Catchment Dominated by Plantation Forestry and Livestock Grazing. Water 8, 436. https://doi.org/10.3390/w8100436.

Kammerbauer, M., 2014. Natural Hazards Governance in Germany, in: Benouar, D. (Ed.), Oxford research encyclopedia of natural hazard science. Oxford University Press, New York, NY.

Kanowski, J., Catterall, C.P., Wardell-Johnson, G.W., 2005. Consequences of broadscale timber plantations for biodiversity in cleared rainforest landscapes of tropical and subtropical Australia. Forest Ecology and Management 208, 359–372. https://doi.org/10.1016/j.foreco.2005.01.018.

Kasperson, R.E., Renn, O., Slovic, P., Brown, H.S., Emel, J., Goble, R., Kasperson, J.X., Ratick, S., 1988. The Social Amplification of Risk: A Conceptual Framework. Risk Analysis 8, 177–187. https://doi.org/10.1111/j.1539-6924.1988.tb01168.x.

Katanha, A., Simatele, D., 2019. Natural hazard mitigation strategies review: Actor-network theory and the eco-based approach understanding in Zimbabwe. Jamba (Potchefstroom, South Africa) 11, 629. https://doi.org/10.4102/jamba.v11i1.629.

Keefer, D.K., 1984. Landslides caused by earthquake. Geological Society of America Bulletin 95. https://doi.org/10.1130/0016-7606(1984)95<406:LCBE>2.0.CO.

Kerns, B.K., Moore, M.M., Timpson, M.E., Hart, S.C., 2003. Soil properties associated with vegetation patterns in a Pinus ponderosa-bunchgrass mosaic. Western North American Naturalist 63, 452–462.

Kimmins, J.P., 2003. Forest Ecology: A Foundation for Sustainable Forest Management and Environmental Ethics in Forestry, 3rd ed. Prentice Hall, Upper Saddle River.

Klaar, M.J., Kidd, C., Malone, E., Bartlett, R., Pinay, G., Chapin, F.S., Milner, A., 2015. Vegetation succession in deglaciated landscapes: implications for sediment and landscape stability. Earth Surf. Process. Landforms 40, 1088–1100. https://doi.org/10.1002/esp.3691.

Klimeš, J., Müllerová, H., Woitsch, J., Bíl, M., Křížová, B., 2020. Century-long history of rural community landslide risk reduction. International Journal of Disaster Risk Reduction 51, 101756. https://doi.org/10.1016/j.ijdrr.2020.101756.

Klinke, A., Renn, O., 1999. Prometheus unbound: Challenges of risk evaluation, risk classification, and risk management. Universität Stuttgart.

Klose, M., Maurischat, P., Damm, B., 2016. Landslide impacts in Germany: A historical and socioeconomic perspective. Landslides 13, 183–199. https://doi.org/10.1007/s10346-015-0643-9.

Korup, O., Seidemann, J., Mohr, C.H., 2019. Increased landslide activity on forested hillslopes following two recent volcanic eruptions in Chile. Nat. Geosci. 12, 284–289. https://doi.org/10.1038/s41561-019-0315-9.

Krauss, J., Bommarco, R., Guardiola, M., Heikkinen, R.K., Helm, A., Kuussaari, M., Lindborg, R., Ockinger, E., Pärtel, M., Pino, J., Pöyry, J., Raatikainen, K.M., Sang, A., Stefanescu, C., Teder, T., Zobel, M., Steffan-Dewenter, I., 2010. Habitat fragmentation causes immediate and time-delayed biodiversity loss at different trophic levels. Ecology Letters 13, 597–605. https://doi.org/10.1111/j.1461-0248.2010.01457.x.

Kreibich, H., Bubeck, P., Kunz, M., Mahlke, H., Parolai, S., Khazai, B., Daniell, J., Lakes, T., Schröter, K., 2014. A review of multiple natural hazards and risks in Germany. Nat Hazards 74, 2279–2304. https://doi.org/10.1007/s11069-014-1265-6.

Kreienkamp, F., Lorenz, P., Tradowsky, J., 2021. Attributionsforschung am DWD – Stand und Pläne. Copernicus Meetings DACH2022-99. Copernicus Meetings. https://meetingorganizer.copernicus.org/DACH2022/DACH2022-99.html.

Kremer, K., Promis, Á., Bauhus, J., 2022. Natural Advance Regeneration of Native Tree Species in Pinus radiata Plantations of South-Central Chile Suggests Potential for a Passive Restoration Approach. Ecosystems 25, 1096–1116. https://doi.org/10.1007/s10021-021-00704-x.

Krueger, E.S., Ochsner, T.E., Engle, D.M., Carlson, J.D., Twidwell, D., Fuhlendorf, S.D., 2015. Soil Moisture Affects Growing-Season Wildfire Size in the Southern Great Plains. Soil Science Society of America Journal 79, 1567–1576. https://doi.org/10.2136/sssaj2015.01.0041.

Kühne, O., Koegst, L., Zimmer, M.-L., Schäffauer, G., 2021. "… Inconceivable, Unrealistic and Inhumane". Internet Communication on the Flood Disaster in West Germany of July 2021 between Conspiracy Theories and Moralization—A Neopragmatic Explorative Study. Sustainability 13, 11427. https://doi.org/10.3390/su132011427.

Kumar, B. Mohan; Nair, P. K. Ramachandran (Hg.) (2011): Carbon Sequestration Potential of Agroforestry Systems: Opportunities and Challenges. Opportunities and challenges. Dordrecht, Heidelberg: Scholars Portal (Advances in agroforestry, 8). Online verfügbar unter http://swb.eblib.com/patron/FullRecord.aspx?p=798930.

La Manna, L., Buduba, C.G., Rostagno, C.M., 2016. Soil erodibility and quality of volcanic soils as affected by pine plantations in degraded rangelands of NW Patagonia. Eur J Forest Res 135, 643–655. https://doi.org/10.1007/s10342-016-0961-z.

La Manna, L., Gaspar, L., Rostagno, C.M., Quijano, L., Navas, A., 2018. Soil changes associated with land use in volcanic soils of Patagonia developed on dynamic landscapes. CATENA 166, 229–239. https://doi.org/10.1016/j.catena.2018.03.025.

La Manna, L., Gaspar, L., Tarabini, M., Quijano, L., Navas, A., 2019. 137Cs inventories along a climatic gradient in volcanic soils of Patagonia: Potential use for assessing medium term erosion processes. CATENA 181, 104089. https://doi.org/10.1016/j.catena. 2019.104089.

Lande, R., DeVries, P.J., Walla, T.R., 2000. When species accumulation curves intersect: implications for ranking diversity using small samples. Oikos 89, 601–605. https://doi. org/10.1034/j.1600-0706.2000.890320.x.

Langdon, B., Pauchard, A., Aguayo, M., 2010. Pinus contorta invasion in the Chilean Patagonia: local patterns in a global context. Biol Invasions 12, 3961–3971. https://doi.org/10. 1007/s10530-010-9817-5.

Larsen, A., Robin, V., Heckmann, T., Fülling, A., Larsen, J.R., Bork, H.-R., 2016. The influence of historic land-use changes on hillslope erosion and sediment redistribution. The Holocene 26, 1248–1261. https://doi.org/10.1177/0959683616638420.

Latorre, J.I., Pedemonte, N.R., 2016. El conflicto forestal en territorio mapuche hoy. Ecología Política 51, 84–87.

Latour, B., 1995. Wir sind nie modern gewesen: Versuch einer symmetrischen Anthropologie. Akademie Verlag, Berlin, 208 pp.

Latour, B., 2014. Existenzweisen: Eine Anthropologie der Modernen. Suhrkamp, Berlin, 666 pp.

Le Bonniec. What is landscape for the Mapuche?

Lennon, J.J., Koleff, P., GreenwooD, J.J.D., Gaston, K.J., 2001. The geographical structure of British bird distributions: diversity, spatial turnover and scale. Journal of Animal Ecology 70, 966–979. https://doi.org/10.1046/j.0021-8790.2001.00563.x.

Licata, J.A., Gyenge, J.E., Fernández, M.E., Schlichter, T.M., Bond, B.J., 2008. Increased water use by ponderosa pine plantations in northwestern Patagonia, Argentina compared with native forest vegetation. Forest Ecology and Management 255, 753–764. https://doi. org/10.1016/j.foreco.2007.09.061.

Lin, Y., Augspurger, C.K., 2008. Impact of spatial heterogeneity of neighborhoods on long-term population dynamics of sugar maple (Acer saccharum). Forest Ecology and Management 255, 3589–3596. https://doi.org/10.1016/j.foreco.2008.02.040.

Loconto, A., Rajão, R., 2020. Governing by models: Exploring the technopolitics of the (in)visilibities of land. Land Use Policy 96, 104241. https://doi.org/10.1016/j.landusepol. 2019.104241.

Lomba, A., Moreira, F., Klimek, S., Jongman, R.H.G., Sullivan, C., Moran, J., Poux, X., Honrado, J.P., Pinto-Correia, T., Plieninger, T., McCracken, D.I., 2020. Back to the future: rethinking socioecological systems underlying high nature value farmlands. Front Ecol Environ 18, 36–42. https://doi.org/10.1002/fee.2116.

Louder, E., Bosak, K., 2019. What the Gringos Brought: Local Perspectives on a Private Protected Area in Chilean Patagonia. Conservat Soc 17, 161. https://doi.org/10.4103/cs.cs_ 17_169.

Luhmann, N., 1991. Soziologie des Risikos. De Gruyter, Berlin, 252 pp.

Maeda, T., Takenaka, H., Warkentin, B.P., 1977. Physical Properties of Allophane Soils, in: Elsevier.

Maintz, J., 2008. Relationalität und räumliche Dynamik von Risiken - ein bioterroristisches Szenario aus der Perspektive der Actor Network Theory, in: Felgentreff, C., Glade, T. (Eds.), Naturrisiken und Sozialkatastrophen. Spektrum Akad. Verl., Berlin.

References

Malkamäki, A., D'Amato, D., Hogarth, N.J., Kanninen, M., Pirard, R., Toppinen, A., Zhou, W., 2018. A systematic review of the socio-economic impacts of large-scale tree plantations, worldwide. Global Environmental Change 53, 90–103. https://doi.org/10.1016/j.gloenvcha.2018.09.001.

Marden, M., 2012. Effectiveness of reforestation in erosion mitigation and implications for future sediment yields, East Coast catchments, New Zealand: A review. New Zealand Geographer 68, 24–35. https://doi.org/10.1111/j.1745-7939.2012.01218.x.

Marin, J., Cortés, J., Aliste, E., Campos, J., 2020. Scientific controversy as a disaster risk factor: The 2007 seismic crisis in Patagonia, Chile. International Journal of Disaster Risk Reduction 49, 101639. https://doi.org/10.1016/j.ijdrr.2020.101639.

Marino, P., Peres, D.J., Cancelliere, A., Greco, R., Bogaard, T.A., 2020. Soil moisture information can improve shallow landslide forecasting using the hydrometeorological threshold approach. Landslides 17, 2041–2054. https://doi.org/10.1007/s10346-020-01420-8.

Markgraf, V., Whitlock, C., Haberle, S., 2007. Vegetation and fire history during the last 18,000 cal yr B.P. in Southern Patagonia: Mallín Pollux, Coyhaique, Province Aisén (45°41'30" S, 71°50'30" W, 640 m elevation). Palaeogeography, Palaeoclimatology, Palaeoecology 254, 492–507. https://doi.org/10.1016/j.palaeo.2007.07.008.

Marston, R.A., 2010. Geomorphology and vegetation on hillslopes: Interactions, dependencies, and feedback loops. Geomorphology 116, 206–217. https://doi.org/10.1016/j.geomorph.2009.09.028.

Martin, D.A., Andrianisaina, F., Fulgence, T.R., Osen, K., Rakotomalala, A.A.N.A., Raveloaritiana, E., Soazafy, M.R., Wurz, A., Andriafanomezantsoa, R., Andriamaniraka, H., Andrianarimisa, A., Barkmann, J., Dröge, S., Grass, I., Guerrero-Ramirez, N., Hänke, H., Hölscher, D., Rakouth, B., Ranarijaona, H.L.T., Randriamanantena, R., Ratsoavina, F.M., Ravaomanarivo, L.H.R., Schwab, D., Tscharntke, T., Zemp, D.C., Kreft, H., 2022. Land-use trajectories for sustainable land system transformations: Identifying leverage points in a global biodiversity hotspot. Proceedings of the National Academy of Sciences of the United States of America 119. https://doi.org/10.1073/pnas.2107747119.

Martínez Pastur, G., Aravena Acuña, M.-C., Silveira, E.M.O., Müller, A. von, La Manna, L., González-Polo, M., Chaves, J.E., Cellini, J.M., Lencinas, M.V., Radeloff, V.C., Pidgeon, A.M., Peri, P.L., 2022. Mapping Soil Organic Carbon Content in Patagonian Forests Based on Climate, Topography and Vegetation Metrics from Satellite Imagery. Remote Sensing 14, 5702. https://doi.org/10.3390/rs14225702.

Mastop, H., Faludi, A., 1997. Evaluation of strategic plans: the performance principle. Environ Plann B Plann Des 24, 815–832. https://doi.org/10.1068/b240815.

Masuda, J.R., Garvin, T., 2006. Place, Culture, and the Social Amplification of Risk. Risk Analysis 26, 437–454. https://doi.org/10.1111/j.1539-6924.2006.00749.x.

McDermott, C.L., Montana, J., Bennett, A., Gueiros, C., Hamilton, R., Hirons, M., Maguire-Rajpaul, V.A., Parry, E., Picot, L., 2023. Transforming the land use governance: Global targets without equity miss the mark. Env Pol Gov 33, 245–257. https://doi.org/10.1002/eet.2027.

McFadden, T.N., Dirzo, R., 2018. Opening the silvicultural toolbox: A new framework for conserving biodiversity in Chilean timber plantations. Forest Ecology and Management 425, 75–84. https://doi.org/10.1016/j.foreco.2018.05.028.

McGarigal, K., Cushman, S., Neel, M., Ene, E., 2002. FRAGSTATS: spatial pattern analysis program for categorical maps.: Technical Report. https://scholar.google.com/citations?user=4zgayliaaaaj&hl=en&oi=sra.

Mendoza, M., 2023. Territorializing Capital: Moreno's Gift and the Political Economy of Nature in Argentine Patagonia, in: Gale-Detrich, T., Ednie, A., Bosak, K. (Eds.), Tourism and Conservation-based Development in the Periphery. Springer International Publishing, Cham, pp. 29–46.

Mendoza, M., Fletcher, R., Holmes, G., Ogden, L.A., Schaeffer, C., 2017. The Patagonian Imaginary: Natural Resources and Global Capitalism at the Far End of the World. Journal of Latin American Geography 16, 93–116. https://doi.org/10.1353/lag.2017.0023.

Metternicht, G., 2018. Land Use and Spatial Planning: Enabling Sustainable Management of Land Resources. Springer International Publishing, Cham, 116 pp.

Ministerio De Agricultura, 1974. Fija Regimen Legal de los Terrenos Forestales o Preferenmente Aptos para la Forestacion, y Establece Normas de Fomento sobre la Materia: Decreto Ley 701.

Molinet, C., Niklitschek, E.J., Coper, S., Diaz, M., Diaz, P.A., Fuentealba, M., Marticorena, F., 2014. Challenges for coastal zoning and sustainable development in the northern Patagonian fjords (Aysen, Chile). lajar 42, 18–29. https://doi.org/10.3856/vol42-issue1-fulltext-2.

Morales Olmos, V., 2022. Forestry and the forest products sector: Production, income and employment, and international trade. Forest Policy and Economics 135, 102648. https://doi.org/10.1016/j.forpol.2021.102648.

Moreira-Dantas, I.R., Söder, M., 2022. Global deforestation revisited: The role of weak institutions. Land Use Policy 122, 106383. https://doi.org/10.1016/j.landusepol.2022.106383.

Morelli, S., Pazzi, V., Nardini, O., Bonati, S., 2022. Framing Disaster Risk Perception and Vulnerability in Social Media Communication: A Literature Review. Sustainability 14, 9148. https://doi.org/10.3390/su14159148.

Morrison, T.H., Adger, W.N., Agrawal, A., Brown, K., Hornsey, M.J., Hughes, T.P., Jain, M., Lemos, M.C., McHugh, L.H., O'Neill, S., van Berkel, D., 2022. Radical interventions for climate-impacted systems. Nat. Clim. Chang. 12, 1100–1106. https://doi.org/10.1038/s41558-022-01542-y.

Morrison, T.H., Adger, W.N., Brown, K., Lemos, M.C., Huitema, D., Phelps, J., Evans, L., Cohen, P., Song, A.M., Turner, R., Quinn, T., Hughes, T.P., 2019. The black box of power in polycentric environmental governance. Global Environmental Change 57, 101934. https://doi.org/10.1016/j.gloenvcha.2019.101934.

Mosquera-Losada, M.R., Santos, M.G.S., Gonçalves, B., Ferreiro-Domínguez, N., Castro, M., Rigueiro-Rodríguez, A., González-Hernández, M.P., Fernández-Lorenzo, J.L., Romero-Franco, R., Aldrey-Vázquez, J.A., Sobrino, C.C., García-Berrios, J.J., Santiago-Freijanes, J.J., 2023. Policy challenges for agroforestry implementation in Europe. Front. For. Glob. Change 6, 1127601. https://doi.org/10.3389/ffgc.2023.1127601.

Mudombi-Rusinamhodzi, G., Thiel, A., 2020. Property rights and the conservation of forests in communal areas in Zimbabwe. Forest Policy and Economics 121, 102315. https://doi.org/10.1016/j.forpol.2020.102315.

Müller-Mahn, D., Everts, J., Stephan, C., 2018. Riskscapes revisited - Exploring the Relationships between risk, space and practice. Erdkunde 72, 197–214.

Municipalidad de Coyhaique, 2011. Plan regulador de Coyhaique.

Murphy, M., Balser, T., Buchmann, N., Hahn, V., Potvin, C., 2008. Linking tree biodiversity to belowground process in a young tropical plantation: Impacts on soil CO_2 flux. Forest Ecology and Management 255, 2577–2588. https://doi.org/10.1016/j.foreco.2008.01.034.

Murphy, M., Danis, W.M., Mack, J., Sayers, J., 2020. From principles to action: Community-based entrepreneurship in the Toquaht Nation. Journal of Business Venturing 35.

Myers, N., Mittermeier, R.A., Mittermeier, C.G., Da Fonseca, G.A., Kent, J., 2000. Biodiversity hotspots for conservation priorities. Nature 403, 853–858. https://doi.org/10.1038/35002501.

Nagendra, H., Ostrom, E., 2012. Polycentric governance of multifunctional forested landscapes. International Journal of the Commons 6, 104–133.

Naranjo, J.A., Arenas, M., Clavero, J., Muñoz, O., 2009. Mass movement-induced tsunamis: main effects during the Patagonian Fjordland seismic crisis in Aisén (45°25'S), Chile. Andean Geology 36, 137–145.

NASA, 1984. Landsat 5 TM: LT52310921984361AAA02. USGS Earth Resources Observation and Science (EROS) Center, NASA Land Processed Distributed Active Archive Center (LP DAAC).

NASA, 2012. Landsat 7 ETM+: LE72320922012014EDC00. USGS Earth Resources Observation and Science (EROS) Center, NASA Land Processed Distributed Active Archive Center (LP DAAC).

Nathan, F., 2008. Risk perception, risk management and vulnerability to landslides in the hill slopes in the city of La Paz, Bolivia. A preliminary statement. Disasters 32, 337–357. https://doi.org/10.1111/j.1467-7717.2008.01043.x.

Neira Ceballos, Z., M. Alarcón, A., Jelves, I., Ovalle, P., Conejeros, A.M., Verdugo, V., 2012. Espacios ecológico-culturales en un Territorio Mapuche de la Región de Araucanía en Chile. Chungará (Arica) 44, 313–323.

Neisser, F.M., 2014. 'Riskscapes' and risk management – Review and synthesis of an actor-network theory approach. Risk Manag 16, 88–120. https://doi.org/10.1057/rm.2014.5.

Newig, J., Lenschow, A., Challies, E., Cotta, B., Schilling-Vacaflor, A., 2019. What is governance in global telecoupling? E&S 24. https://doi.org/10.5751/ES-11178-240326.

Nilsson, M., Persson, Å., 2012. Can Earth system interactions be governed? Governance functions for linking climate change mitigation with land use, freshwater and biodiversity protection. Ecological Economics 75, 61–71. https://doi.org/10.1016/j.ecolecon.2011.12.015.

North, M.P., Stevens, J.T., Greene, D.F., Coppoletta, M., Knapp, E.E., Latimer, A.M., Restaino, C.M., Tompkins, R.E., Welch, K.R., York, R.A., Young, D.J., Axelson, J.N., Buckley, T.N., Estes, B.L., Hager, R.N., Long, J.W., Meyer, M.D., Ostoja, S.M., Safford, H.D., Shive, K.L., Tubbesing, C.L., Vice, H., Walsh, D., Werner, C.M., Wyrsch, P., 2019. Tamm Review: Reforestation for resilience in dry western U.S. forests. Forest Ecology and Management 432, 209–224. https://doi.org/10.1016/j.foreco.2018.09.007.

Oberdorfer, E., 1960. Pflanzensoziologische Studien in Chile. Cramer, Weinheim.

Ogden, C., J., B., J., Stretton, K., Anderson, S., 1997. Plant species richness under Pinus radiata stands on the central North Island volcanic plateau, New Zealand. New Zealand Journal of Ecology 21, 17–29.

Oksanen, J., Blanchet, G.F., Friendly, M., Kindt, R., Legendre, P., McGlinn, D., R. Minchin , Peter R., O'Hara, R.B., Simpson, G.L., Solymos, P., Stevens, M.H.H., Szoecs, E., Wagner, H., 2019. vegan: Community Ecology Package: R package version 2.5–6.

Olivares-Contreras, V.A., Mattar, C., Gutiérrez, A.G., Jiménez, J.C., 2019. Warming trends in Patagonian subantartic forest. International Journal of Applied Earth Observation and Geoinformation 76, 51–65. https://doi.org/10.1016/j.jag.2018.10.015.

Oliver-Smith, A., 2013. A matter of choice. International Journal of Disaster Risk Reduction 3, 1–3. https://doi.org/10.1016/j.ijdrr.2012.12.001.

Olofsson, A., Öhman, S., Nygren, K.G., 2016. An intersectional risk approach for environmental sociology. Environmental Sociology 2, 346–354. https://doi.org/10.1080/23251042.2016.1246086.

Oppikofer, T., Hermanns, R.L., Redfield, T.F., Sepúlveda, S.A., Duhart, P., Bascuñán, I., 2012. Morphologic description of the Punta Cola rock avalanche and associated minor rockslides caused by 21 April 2007 Aysén earthquake (Patagonia, Southern Chile). Revista de la Asociación Geológica Argentina, 339–353.

Ortiz, J., Dube, F., Neira, P., Hernández Valera, R.R., Souza Campos, P.M. de, Panichini, M., Pérez-San Martín, A., Stolpe, N.B., Zagal, E., Curaqueo, G., 2023. Comparative Study between Silvopastoral and Agroforest Systems on Soil Quality in a Disturbed Native Forest of South-Central Chile. Agronomy 13, 2683. https://doi.org/10.3390/agronomy13112683.

Ossandón, A.O., Vásquez, F.A., Alba, D.M., 2020. Ordenamiento territorial en Chile: Nuevo escenario para la gobernanza regional. Rev. geogr. Norte Gd., 31–49. https://doi.org/10.4067/S0718-34022020000300031.

Ostrom, E., 1990. Governing the Commons. Cambridge University Press.

Ostrom, E., 1994. Institutional Analysis, Design Principles and Threats to Sustainalbe Community Governance and Management of Commons, in: Pomeroy, R.S. (Ed.), Community Management and Common Property of Coastal Fisheries in Asia and the Pacific: Concepts, Methods and Experiences, pp. 34–50.

Ovalle, C., Avendaño, J., Aronson, J., Del Pozo, A., 1996. Land occupation patterns and vegetation structure in the anthropogenic savannas (espinales) of central Chile. Forest Ecology and Management 86, 129–139. https://doi.org/10.1016/S0378-1127(96)03786-3.

Ovalle, L.R., 2011. Ocupación y Desarollo: de la Provincia de Aisén, Coyhaique.

Page, M., Trustrum, N., Gomez, B., 2000. Implications of a Century of Anthropogenic Erosion for Future Land Use in the Gisborne-East Coast Region of New Zealand. New Zealand Geographer 56, 13–24. https://doi.org/10.1111/j.1745-7939.2000.tb01571.x.

Parisien, M.-A., Barber, Q.E., Hirsch, K.G., Stockdale, C.A., Erni, S., Wang, X., Arseneault, D., Parks, S.A., 2020. Fire deficit increases wildfire risk for many communities in the Canadian boreal forest. Nat Commun 11, 2121. https://doi.org/10.1038/s41467-020-15961-y.

Paritsis, J., Aizen, M.A., 2008. Effects of exotic conifer plantations on the biodiversity of understory plants, epigeal beetles and birds in Nothofagus dombeyi forests. Forest Ecology and Management 255, 1575–1583. https://doi.org/10.1016/j.foreco.2007.11.015.

Paritsis, J., Veblen, T.T., Holz, A., 2015. Positive fire feedbacks contribute to shifts from Nothofagus pumilio forests to fire-prone shrublands in Patagonia. J Veg Sci 26, 89–101. https://doi.org/10.1111/jvs.12225.

References

Parizek, B., Rostagno, C.M., Sottini, R., 2002. Soil Erosion as Affected by Shrub Encroachment in Northeastern Patagonia. Journal of Range Management 55, 43. https://doi.org/10.2307/4003261.

Pauchard, A., Aguayo, M., PEÑA, E., Urrutia, R., 2006. Multiple effects of urbanization on the biodiversity of developing countries: The case of a fast-growing metropolitan area (Concepción, Chile). Biological Conservation 127, 272–281. https://doi.org/10.1016/j.biocon.2005.05.015.

Pauchard, A., Alaback, P.B., 2004. Influence of Elevation, Land Use, and Landscape Context on Patterns of Alien Plant Invasions along Roadsides in Protected Areas of South-Central Chile. Conservation Biology 18, 238–248. https://doi.org/10.1111/j.1523-1739.2004.00300.x.

Pauchard, A., García, R.A., PEÑA, E., González, C., Cavieres, L.A., Bustamante, R.O., 2008. Positive feedbacks between plant invasions and fire regimes: Teline monspessulana (L.) K. Koch (Fabaceae) in central Chile. Biol Invasions 10, 547–553. https://doi.org/10.1007/s10530-007-9151-8.

Pawlik, Ł., 2013. The role of trees in the geomorphic system of forested hillslopes—A review. Earth-Science Reviews 126, 250–265. https://doi.org/10.1016/j.earscirev.2013.08.007.

Pawlik, Ł., Migoń, P., Owczarek, P., Kacprzak, A., 2013. Surface processes and interactions with forest vegetation on a steep mudstone slope, Stołowe Mountains, SW Poland. CATENA 109, 203–216. https://doi.org/10.1016/j.catena.2013.03.011.

Pawlik, Ł., Phillips, J.D., Šamonil, P., 2016. Roots, rock, and regolith: Biomechanical and biochemical weathering by trees and its impact on hillslopes—A critical literature review. Earth-Science Reviews 159, 142–159. https://doi.org/10.1016/j.earscirev.2016.06.002.

Pawlik, Ł., Šamonil, P., 2018. Soil creep: The driving factors, evidence and significance for biogeomorphic and pedogenic domains and systems – A critical literature review. Earth-Science Reviews 178, 257–278. https://doi.org/10.1016/j.earscirev.2018.01.008.

Pawson, S., Brockerhoff, E.G., Didham, R., 2009. Native forest generalists dominate carabid assemblages along a stand age chronosequence in an exotic Pinus radiata plantation. Forest Ecology and Management.

Pedersen, K.L. A landscape forensic investigation of the 2021 flood in the Ahr Valley, Germany.

Pellissier, V., Gallet, S., Rozé, F., 2004. Comparison of the vegetation and seed bank on hedge banks of different ages in Brittany, France. Environmental Management 34, 52–61. https://doi.org/10.1007/s00267-004-0041-x.

Pérez, C., Simonetti, J.A., 2022. Subsidy Accountability and Biodiversity Loss Drivers: Following the Money in the Chilean Silvoagricultural Sector. Sustainability 14, 15411. https://doi.org/10.3390/su142215411.

Pérez, P., Walter, D., 2020. Beyond the "desert". Indigenous genocide as a structuring event in Northern Patagonia, in: Larson, C. (Ed.), The Conquest of the Desert: Argentina's Indigenous Peoples and the Battle for History. University of New Mexico Press; Project MUSE, Albuquerque, Baltimore, Md.

Pérez Piñán, A., Vibert, E., 2019. The View from the Farm: Gendered Contradictions of the Measurement Imperative in Global Goals. Journal of Human Development and Capabilities 20, 436–450. https://doi.org/10.1080/19452829.2019.1659237.

Peri, P.L., Hansen, N.E., Bahamonde, H.A., Lencinas, M.V., Müller, A.R. von, Ormaechea, S., Gargaglione, V., Soler, R., Tejera, L.E., Lloyd, C.E., Pastur, G.M., 2016. Silvopastoral Systems Under Native Forest in Patagonia Argentina, in: Peri, P.L., Dube, F., Varella, A. (Eds.), Silvopastoral Systems in Southern South America, 1st ed. Springer International Publishing; Springer e-books; Imprint; Springer, Cham, pp. 117–168.

Peri, P.L., Nahuelhual, L., Martínez Pastur, G., 2021. Ecosystem Services as a Tool for Decision-Making in Patagonia, in: Peri, P.L., Pastur, G.M., Nahuelhual, L. (Eds.), Ecosystem Services in Patagonia: A Multi-Criteria Approach for an Integrated Assessment. Springer, Cham.

Perley, C., 1994. Biodiversity, Sustainability and a Land Ethic. New Zealand Forestry 39.

Peters, L.E., 2021. Beyond disaster vulnerabilities: An empirical investigation of the causal pathways linking conflict to disaster risks. International Journal of Disaster Risk Reduction, 102092. https://doi.org/10.1016/j.ijdrr.2021.102092.

Phillips, C., Marden, M., 2006. Reforestation Schemes to Manage Regional Landslide Risk, in: Glade, T., Anderson, M.G., Crozier, M.J. (Eds.), Landslide hazard and risk. Wiley, Chichester, pp. 517–542.

Phillips, C.J., Rey, F., Marden, M., Liébault, F., 2013. Revegetation of steeplands in France and New Zealand: geomorphic and policy responses. N.Z. j. of For. Sci. 43, 14. https://doi.org/10.1186/1179-5395-43-14.

Plieninger, T., Bieling, C. (Eds.), 2012. Resilience and the cultural landscape: Understanding and managing change in human-shaped environments. Cambridge University Press, Cambridge, UK, New York, 348 pp.

Plieninger, T., Muñoz-Rojas, J., Buck, L.E., Scherr, S.J., 2020. Agroforestry for sustainable landscape management. Sustain Sci 15, 1255–1266. https://doi.org/10.1007/s11625-020-00836-4.

Plieninger, T., Shamohamadi, S., García-Martín, M., Quintas-Soriano, C., Shakeri, Z., Valipour, A., 2023. Community, pastoralism, landscape: Eliciting values and human-nature connectedness of forest-related people. Landscape and Urban Planning 233, 104706. https://doi.org/10.1016/j.landurbplan.2023.104706.

Pohl, J., 2008. Die Entstehung der geographischen Hazardforschung, in: Felgentreff, C., Glade, T. (Eds.), Naturrisiken und Sozialkatastrophen. Spektrum Akad. Verl., Berlin.

Pohl, L., Maurischat, P., Schiedung, M., Weber, T.K.D., Winkler, P., 2022. Soil science challenges—An interdisciplinary overview of current and future topics. J. Plant Nutr. Soil Sci. 185, 691–693. https://doi.org/10.1002/jpln.202200396.

Poliseli, L., Coutinho, J.G.E., Viana, B., Russo, F., El-Hani, C.N., 2022. Philosophy of science in practice in ecological model building. Biol Philos 37, 1–27. https://doi.org/10.1007/s10539-022-09851-4.

Polo Blanco, J., Piñeiro Aguiar, E., 2019. Ciencia moderna, planeta torturado Una reflexión crítica sobre el modo eurocéntrico de conocer la naturaleza e intervenir en el medio ambiente. Izquierdas (Santiago), 194–217. https://doi.org/10.4067/S0718-50492019000200194.

Popp, A., Humpenöder, F., Weindl, I., Bodirsky, B.L., Bonsch, M., Lotze-Campen, H., Müller, C., Biewald, A., Rolinski, S., Stevanovic, M., Dietrich, J.P., 2014. Land-use protection for climate change mitigation. Nat. Clim. Chang. 4, 1095–1098. https://doi.org/10.1038/nclimate2444.

Popp, A., Krause, M., Dietrich, J.P., Lotze-Campen, H., Leimbach, M., Beringer, T., Bauer, N., 2012. Additional CO2 emissions from land use change—Forest conservation as a precondition for sustainable production of second generation bioenergy. Ecological Economics 74, 64–70. https://doi.org/10.1016/j.ecolecon.2011.11.004.

Porta, J., López-Acevedo M., Poch, R.M., 2013. Edafología. Uso y protección de suelos. Munid-Prensa.

Porter, J., Demeritt, D., 2012. Flood-Risk Management, Mapping, and Planning: The Institutional Politics of Decision Support in England. Environ Plan A 44, 2359–2378. https://doi.org/10.1068/a44660.

Potton, C., 1994. A public perception of plantation forestry. New Zealand Forestry 39.

Puppim de Oliveira, Jose A., Fra.Paleo, U., 2016. Lost in participation: How local knowledge was overlooked in land use planning and risk governance in Tōhoku, Japan. Land Use Policy 52, 543–551. https://doi.org/10.1016/j.landusepol.2014.09.023.

Quijano, A., 2000. Coloniality of Power and Eurocentrism in Latin America. International Sociology 15, 215–232. https://doi.org/10.1177/0268580900015002005.

R Development Core Team, 2008. R: A language and environment for statistical computing. R Foundation for Statistical Computing, Vienna, Austria.

Ramcilovic-Suominen, S., Kröger, M., Dressler, W., 2022. From pro-growth and planetary limits to degrowth and decoloniality: An emerging bioeconomy policy and research agenda. Forest Policy and Economics 144, 102819. https://doi.org/10.1016/j.forpol.2022.102819.

Ramsey, M.M., Muñoz-Erickson, T.A., Mélendez-Ackerman, E., Nytch, C.J., Branoff, B.L., Carrasquillo-Medrano, D., 2019. Overcoming barriers to knowledge integration for urban resilience: A knowledge systems analysis of two-flood prone communities in San Juan, Puerto Rico. Environmental Science & Policy 99, 48–57. https://doi.org/10.1016/j.envsci.2019.04.013.

Ranjan, R., Castillo, A., Morales, K., 2021a. Mapuche cosmovision and territorial rights: An interdisciplinary approach to understand the conflict of Wallmapu, Chile. rupkatha 13. https://doi.org/10.21659/rupkatha.v13n1.17.

Ranjan, R., Castillo, A., Morales, K., 2021b. Mapuche cosmovision and territorial rights: An interdisciplinary approach to understand the conflict of Wallmapu, Chile. rupkatha 13. https://doi.org/10.21659/rupkatha.v13n1.17.

Rasmussen, M.B., Figueroa, L., 2022. Patagonian ground rules: institutionalizing access at the frontier. The Journal of Peasant Studies, 1–20. https://doi.org/10.1080/03066150.2021.2009461.

Rastandeh, A., Jarchow, M. (Eds.), 2022a. Creating Resilient Landscapes in an Era of Climate Change: Global Case Studies and Real-World Solutions. Routledge, [Erscheinungsort nicht ermittelbar], 1 Online-Ressource.

Rastandeh, A., Jarchow, M., 2022b. Landscape Resilience in the face of climate change, in: Rastandeh, A., Jarchow, M. (Eds.), Creating Resilient Landscapes in an Era of Climate Change: Global Case Studies and Real-World Solutions. Routledge, [Erscheinungsort nicht ermittelbar], pp. 1–10.

Reinstädtler, S., 2023. Transboundary systems provision for sustainable and resilient climate risk-, disaster risk- and crisis management in the flood disaster-damaged Ahr Valley in Germany - Implementing Spree Forest and Lusatia Regions Land and Environmental Systems Model.

Renck, V., Ludwig, D., Bollettin, P., Reis-Filho, J.A., Poliseli, L., El-Hani, C., 2023. Taking fishers' knowledge and its implications to fisheries policy seriously. E&S 28. https://doi.org/10.5751/ES-14104-280207.

Renn, O., Klinke, A., 2016. Risk Perception and Its Impacts on Risk Governance, in: Shugart, H.H. (Ed.), Oxford research encyclopedia of environmental science: Oxford research encyclopedias. Oxford University Press, New York.

Renn, O., Schweizer, P.-J., Dreyer, M., Klinke, A., 2007. Risiko: Über den gesellschaftlichen Umgang mit Unsicherheit. oekom verlag Gesellschaft für ökologische Kommunikation mbH, München, 271 pp.

Richards, P., 2010. Of Indians and Terrorists: How the State and Local Elites Construct the Mapuche in Neoliberal Multicultural Chile. J. Lat. Am. Stud. 42, 59–90. https://doi.org/10.1017/S0022216X10000052.

Richardson, K., Steffen, W., Lucht, W., Bendtsen, J., Cornell, S.E., Donges, J.F., Drüke, M., Fetzer, I., Bala, G., Bloh, W. von, Feulner, G., Fiedler, S., Gerten, D., Gleeson, T., Hofmann, M., Huiskamp, W., Kummu, M., Mohan, C., Nogués-Bravo, D., Petri, S., Porkka, M., Rahmstorf, S., Schaphoff, S., Thonicke, K., Tobian, A., Virkki, V., Wang-Erlandsson, L., Weber, L., Rockström, J., 2023. Earth beyond six of nine planetary boundaries. Science advances 9, eadh2458. https://doi.org/10.1126/sciadv.adh2458.

Rickli, C., Bebi, P., Graf, F., Moos, C., 2019. Shallow Landslides: Retrospective Analysis of the Protective Effects of Forest and Conclusions for Prediction, in: Wu, W. (Ed.), Recent Advances in Geotechnical Research. Springer International Publishing, Cham, pp. 175–185.

Rivera, A., Bown, F., 2013. Recent glacier variations on active ice capped volcanoes in the Southern Volcanic Zone (37°–46°S), Chilean Andes. Journal of South American Earth Sciences 45, 345–356. https://doi.org/10.1016/j.jsames.2013.02.004.

Robbins, P., Maddock, T., 2000. Interrogating Land Cover Categories: Metaphor and Method in Remote Sensing. Cartography and Geographic Information Science 27, 295–309. https://doi.org/10.1559/152304000783547740.

Rocha, J.C., Mazzeo, N., Piaggio, M., Carriquiry, M., 2020. Seeking sustainable pathways for land use in Latin America. E&S 25. https://doi.org/10.5751/ES-11824-250317.

Rockström, J., Beringer, T., Hole, D., Griscom, B., Mascia, M.B., Folke, C., Creutzig, F., 2021. Opinion: We need biosphere stewardship that protects carbon sinks and builds resilience. Proceedings of the National Academy of Sciences of the United States of America 118. https://doi.org/10.1073/pnas.2115218118.

Rojo-Mendoza, F., Salinas-Silva, C., Alvarado-Peterson, V., 2022. The end of indigenous territory? Projected counterurbanization in rural Protected Indigenous Areas in Temuco, Chile. Geoforum 133, 66–78. https://doi.org/10.1016/j.geoforum.2022.05.012.

Rosa, M. de, 2018. Land Use and Land-use Changes in Life Cycle Assessment: Green Modelling or Black Boxing? Ecological Economics 144, 73–81. https://doi.org/10.1016/j.ecolecon.2017.07.017.

Rostagno, C., Coronato, F., Valle, H.D., Puebla, D., 1999. Runoff and Erosion in Five Land Units of a Closed Basin of Northeastern Patagonia. Arid Soil Research and Rehabilitation 13, 281–292. https://doi.org/10.1080/089030699263311.

Rothe, D., 2017. Seeing like a satellite: Remote sensing and the ontological politics of environmental security. Security Dialogue 48, 334–353. https://doi.org/10.1177/0967010617709399.

References

RStudio Team, 2019. RStudio: Integrated Development Environment for R. RStudio, Inc., Boston, MA, Boston, MA.

Rudel, T.K., Meyfroidt, P., 2014. Organizing anarchy: The food security–biodiversity–climate crisis and the genesis of rural land use planning in the developing world. Land Use Policy 36, 239–247. https://doi.org/10.1016/j.landusepol.2013.07.008.

Ruiz Agudelo, C.A., Mazzeo, N., Díaz, I., Barral, M.P., Piñeiro, G., Gadino, I., Roche, I., Acuña-Posada, R.J., 2020. Land use planning in the Amazon basin: challenges from resilience thinking. 1708–3087 25. https://doi.org/10.5751/ES-11352-250108.

Saavedra, B., Simonetti, J.A., 2005. Small mammals of Maulino forest remnants, a vanishing ecosystem of south-central Chile. 1864–1547 69, 337–348. https://doi.org/10.1515/mamm.2005.027.

Sager, T., 2022. Advocacy planning: were expectations fulfilled? Planning Perspectives 37, 1205–1230. https://doi.org/10.1080/02665433.2022.2040189.

Salazar, G., Riquelme Maulén, W., 2022. The Space-Time Compression of Indigenous Toponomy: The Case of Mapuche Toponomy in Chilean Northpatagonia. Geographical Review 112, 641–666. https://doi.org/10.1080/00167428.2020.1839898.

Salgado, H., Bailey, J., Tiller, R., Ellis, J., 2015. Stakeholder perceptions of the impacts from salmon aquaculture in the Chilean Patagonia. Ocean & Coastal Management 118, 189–204. https://doi.org/10.1016/j.ocecoaman.2015.07.016.

Salt, D., Lindenmayer, D., Hobbs, R.J., 2004. Trees and biodiversity: A guide for Australian farm forestry. RIRDC, Barton, A.C.T., 201 pp.

Sánchez-Jardón, L., Acosta, B., Del Pozo, A., Casado, M.A., Ovalle, C., Elizalde, H.F., Hepp, C., Miguel, J.M. de, 2010. Grassland productivity and diversity on a tree cover gradient in Nothofagus pumilio in NW Patagonia. Agriculture, Ecosystems & Environment 137, 213–218. https://doi.org/10.1016/j.agee.2010.02.006.

Sanderson, E.W., Jaiteh, M., Levy, M.A., Redford, K.H., Wannebo, A.V., Woolmer, G., 2002. The Human Footprint and the Last of the Wild. BioScience 52, 891. https://doi.org/10.1641/0006-3568(2002)052[0891:THFATL]2.0.CO;2.

Sandoval, V., Wisner, B., Voss, M., 2014. Natural Hazards Governance in Chile, in: Benouar, D. (Ed.), Oxford research encyclopedia of natural hazard science. Oxford University Press, New York, NY.

Sandoval E., M., 2014. Propiedades físicas de los souelos de la Región Aysén, in: Hepp, C.K., Stolpe, N.B. (Eds.), Caracterización y Propiedades de los Suelos de la Patagonia Occidental (Aysén), 300th ed. Imprenta América, Temuco, Chile.

Sandoval E., M., Dörner F., J., Seguel S., O., Cuevas B., J., Rivera S., D., 2012. Métodos de análisis físicos de suelos, 5th ed. Publicaciones Departamento de Suelos y Recursos Naturales, Chillan.

Sapountzaki, K., Wanczura, S., Casertano, G., Greiving, S., Xanthopoulos, G., Ferrara, F.F., 2011. Disconnected policies and actors and the missing role of spatial planning throughout the risk management cycle. Nat Hazards 59, 1445–1474. https://doi.org/10.1007/s11069-011-9843-3.

Saunders, W., Becker, J.S., 2015. A discussion of resilience and sustainability: Land use planning recovery from the Canterbury earthquake sequence, New Zealand. International Journal of Disaster Risk Reduction 14, 73–81. https://doi.org/10.1016/j.ijdrr.2015.01.013.

Schaldach, R., Göpel, J., Schüngel, J., 2020. An Integrated Modelling Approach for Land Use Changes on Different Scales. Landscape Modelling and Decision Support, 509–524. https://doi.org/10.1007/978-3-030-37421-1_26.

Scheffer, F., Schachtschabel, P., Blume, H.-P., Brümmer, G.W., Horn, R., Kandeler, E., Kögel-Knabner, I., Kretzschmar, R., Stahr, K., Thiele-Bruhn, S., Welp, G., Wilke, B.-M., 2010. Lehrbuch der Bodenkunde, 16th ed. Spektrum Akademischer Verlag, Heidelberg.

Schils, R.L., Bufe, C., Rhymer, C.M., Francksen, R.M., Klaus, V.H., Abdalla, M., Milazzo, F., Lellei-Kovács, E., Berge, H. ten, Bertora, C., Chodkiewicz, A., Dămătîrcă, C., Feigenwinter, I., Fernández-Rebollo, P., Ghiasi, S., Hejduk, S., Hiron, M., Janicka, M., Pellaton, R., Smith, K.E., Thorman, R., Vanwalleghem, T., Williams, J., Zavattaro, L., Kampen, J., Derkx, R., Smith, P., Whittingham, M.J., Buchmann, N., Price, J.P.N., 2022. Permanent grasslands in Europe: Land use change and intensification decrease their multifunctionality. Agriculture, Ecosystems & Environment 330, 107891. https://doi.org/10.1016/j.agee.2022.107891.

Schirmer, J., 2007. Plantations and social conflict: exploring the differences between smallscale and large-scale plantation forestry. Small-scale Forestry 6, 19–33. https://doi.org/10.1007/s11842-007-9001-7.

Schmalz, S., Graf, J., Julián-Vejar, D., Sittel, J., Alister Sanhueza, P., 2023. Challenging the three faces of extractivism: the Mapuche struggle and the forestry industry in Chile. Globalizations 20, 365–383. https://doi.org/10.1080/14747731.2022.2091867.

Schmidt, C., 2022. Landscape Resilience: Basics, Case Studies, Practical Recommendations, 1st ed. Springer Berlin Heidelberg; Imprint Springer, Berlin, Heidelberg, 245 pp.

Schmithüsen, J., 1956. Die räumliche Ordnung der chilenischen Vegetation. Stollfuss.

Schneiderbauer, S., Fontanella Pisa, P., Delves, J.L., Pedoth, L., Rufat, S., Erschbamer, M., Thaler, T., Carnelli, F., Granados-Chahin, S., 2021. Risk perception of climate change and natural hazards in global mountain regions: A critical review. The Science of the total environment 784, 146957. https://doi.org/10.1016/j.scitotenv.2021.146957.

Schröder, P., 2023. Limits and beyond: 50 years on from The limits to growth , what did we learn and what's next? International Affairs 99, 1342–1343. https://doi.org/10.1093/ia/iia d085.

Schulz, M., Mack, B., Renn, O. (Eds.), 2012. Fokusgruppen in der empirischen Sozialwissenschaft: Von der Konzeption bis zur Auswertung. Springer VS, Wiesbaden, 209 pp.

Schütze, N., Thiel, A., in review. How and why do actors in polycentric water governance coordinate or not? A comparative case study on the modernization of irrigation on Spain.

Selwyn, B., 2023. De/Recolonising Development: Fanon, Rostow, and the Violence of Social Change. Antipode, anti.12944. https://doi.org/10.1111/anti.12944.

Senbeta, F., Teketay, D., 2001. Regeneration of indigenous woody species under the canopies of tree plantations in Central Ethiopia. Tropical Ecology.

Senn, J.A., Fassnacht, F.E., Eichel, J., Seitz, S., Schmidtlein, S., 2020. A new concept for estimating the influence of vegetation on throughfall kinetic energy using aerial laser scanning. Earth Surf. Process. Landforms 45, 1487–1498. https://doi.org/10.1002/esp.4820.

Seppelt, R., Lautenbach, S., Volk, M., 2013. Identifying trade-offs between ecosystem services, land use, and biodiversity: a plea for combining scenario analysis and optimization on different spatial scales. Current Opinion in Environmental Sustainability 5, 458–463. https://doi.org/10.1016/j.cosust.2013.05.002.

Sepúlveda, B., Guyot, S., 2016. Escaping the Border, Debordering the Nature: Protected Areas, Participatory Management, and Environmental Security in Northern Patagonia (i.e. Chile and Argentina). Globalizations 13, 767–786. https://doi.org/10.1080/147 47731.2015.1133045.

Sepúlveda, S.A., Serey, A., Lara, M., Pavez, A., Rebolledo, S., 2010. Landslides induced by the April 2007 Aysén Fjord earthquake, Chilean Patagonia. Landslides 7, 483–492. https://doi.org/10.1007/s10346-010-0203-2.

SEREMI MMA-AYSÉN, 2018. Estrategia Regional de Biodiversidad: Región de Aysén del General Carlos Ibáñez del Campo, 80 pp.

Sidle, R.C., Bogaard, T.A., 2016. Dynamic earth system and ecological controls of rainfall-initiated landslides. Earth-Science Reviews 159, 275–291. https://doi.org/10.1016/j.earscirev.2016.05.013.

Sidle, R.C., Ochiai, H., 2006. Landslides: Processes, prediction, and land use. American Geophysical Union, Washington, DC, 312 pp.

Sieber, I.M., Cybele, C., 2024. Knowledge Co-Creation for Enhanced Ecosystem Services Management on Islands. Copernicus Meetings EGU24–19975. Copernicus Meetings. https://meetingorganizer.copernicus.org/EGU24/EGU24-19975.html.

Simberloff, D., Nuñez, M.A., Ledgard, N.J., Pauchard, A., Richardson, D.M., Sarasola, M., van Wilgen, B.W., ZALBA, S.M., ZENNI, R.D., BUSTAMANTE, R., PEÑA, E., Ziller, S.R., 2010. Spread and impact of introduced conifers in South America: Lessons from other southern hemisphere regions. Austral Ecology 35, 489–504. https://doi.org/10.1111/j.1442-9993.2009.02058.x.

Simonetti, J.A., Grez, A.A., Celis-Diez, J.L., Bustamante, R.O., 2007. Herbivory and seedling performance in a fragmented temperate forest of Chile. Acta Oecologica 32, 312–318. https://doi.org/10.1016/j.actao.2007.06.001.

Simpson, E.H., 1949. Measurement of Diversity. Nature 163, 688. https://doi.org/10.1038/163688a0.

Slovic, P., 1987. Perception of Risk. Science, 280–285.

Slovic, P., Finucane, M.L., Peters, E., MacGregor, D.G., 2007. The affect heuristic. European Journal of Operational Research 177, 1333–1352. https://doi.org/10.1016/j.ejor.2005.04.006.

Slovic, P., Fischhoff, B., Lichtenstein, S., 1984. Behavioral decision theory perspectives on risk and safety. Acta Psychologica 56, 183–203. https://doi.org/10.1016/0001-6918(84)90018-0.

Slovic, P., Kunreuther, H., White, G., 2011. Decision Processes, Rationality and Adjustment to Natural Hazards, in: Slovic, P. (Ed.), The perception of risk. Earthscan, London.

Smith, A., Stirling, A., 2008. Social-ecological resilience and socio-technical transitions: critical issues for sustainability governance. STEPS Centre.

Smith, B., Wilson, J.B., 1996. A Consumer's Guide to Evenness Indices. Oikos 76, 70. https://doi.org/10.2307/3545749.

Smith-Ramírez, C., 2004. The Chilean coastal range: a vanishing center of biodiversity and endemism in South American temperate rainforests. Biodivers Conserv 13, 373–393. https://doi.org/10.1023/B:BIOC.0000006505.67560.9f.

Somos-Valenzuela, M.A., Oyarzún-Ulloa, J.E., Fustos-Toribio, I.J., Garrido-Urzua, N., Chen, N., 2020a. The mudflow disaster at Villa Santa Lucía in Chilean Patagonia: understandings and insights derived from numerical simulation and postevent field surveys.

Nat. Hazards Earth Syst. Sci. 20, 2319–2333. https://doi.org/10.5194/nhess-20-2319-2020.

Somos-Valenzuela, M.A., Oyarzún-Ulloa, J.E., Fustos-Toribio, I.J., Garrido-Urzua, N., Ningsheng, C., 2020b. Hidden Hazards: the conditions that potentially enabled the mudflow disaster at Villa Santa Lucía in Chilean Patagonia. Nat. Hazards Earth Syst. Sci. 20, e2019–419. https://doi.org/10.5194/nhess-2019-419.

Sosa, N., Torres, A., Castro-Lopez, V., Velázquez, A., 2023. Participatory Landscape Conservation: A Case Study of a Seasonally Dry Tropical Forest in Michoacan, Mexico. Land 12, 2016. https://doi.org/10.3390/land12112016.

Soto, M.-V., Arriagada-González, J., Molina-Benavides, M., Cabello, M., Contreras-Alonso, M., Ibarra, I., Guevara, G., Sepúlveda, S.A., Maerker, M., 2023. Geodynamic 'Hotspots' in a Periglacial Landscape: Natural Hazards and Impacts on Productive Activities in Chilean Fjordlands, Northern Patagonia. Geosciences 13, 209. https://doi.org/10.3390/geosciences13070209.

Sotomayor, A., Schmidt, H., Salinas, J., Schmidt, A., Sánchez-Jardón, L., Alonso, M., 2016. Silvopastoral Systems in the Aysén and Magallanes Regions of the Chilean Patagonia, in: Peri, P.L., Dube, F., Varella, A. (Eds.), Silvopastoral Systems in Southern South America, 1st ed. Springer International Publishing; Springer e-books; Imprint; Springer, Cham, pp. 213–230.

Soule, B., 2012. Coupled seismic and socio-political crises: The case of Puerto Aysen in 2007. Journal of Risk Research 15, 21–37. https://doi.org/10.1080/13669877.2011.591498.

Späth, F., Rajtschan, V., Weber, T.K.D., Morandage, S., Lange, D., Abbas, S.S., Behrendt, A., Ingwersen, J., Streck, T., Wulfmeyer, V., 2023. The land–atmosphere feedback observatory: a new observational approach for characterizing land–atmosphere feedback. Geosci. Instrum. Method. Data Syst. 12, 25–44. https://doi.org/10.5194/gi-12-25-2023.

Stallins, J.A., 2006. Geomorphology and ecology: Unifying themes for complex systems in biogeomorphology. Geomorphology 77, 207–216. https://doi.org/10.1016/j.geomorph.2006.01.005.

Stanchi, S., D'Amico, M., Zanini, E., Freppaz, M., 2015. Liquid and plastic limits of mountain soils as a function of the soil and horizon type. CATENA 135, 114–121. https://doi.org/10.1016/j.catena.2015.07.021.

Stanchi, S., Freppaz, M., Godone, D., Zanini, E., 2013. Assessing the susceptibility of alpine soils to erosion using soil physical and site indicators. Soil Use Manage 29, 586–596. https://doi.org/10.1111/sum.12063.

Steger, S., Mair, V., Kofler, C., Pittore, M., Zebisch, M., Schneiderbauer, S., 2021. Correlation does not imply geomorphic causation in data-driven landslide susceptibility modelling - Benefits of exploring landslide data collection effects. The Science of the total environment 776, 145935. https://doi.org/10.1016/j.scitotenv.2021.145935.

Stephens, S.S., Wagner, M.R., 2007. Forest Plantations and Biodiversity: A Fresh Perspective. j for 105, 307–313. https://doi.org/10.1093/jof/105.6.307.

Stine, M.B., 2016. Biogeomorphic disturbance: A case study on associations and methods after fire within the alpine treeline ecotone. CATENA 145, 107–117. https://doi.org/10.1016/j.catena.2016.06.001.

Stokes, A., Douglas, G.B., Fourcaud, T., Giadrossich, F., Gillies, C., Hubble, T., Kim, J.H., Loades, K.W., Mao, Z., McIvor, I.R., Mickovski, S.B., Mitchell, S., Osman, N., Phillips,

C., Poesen, J., Polster, D., Preti, F., Raymond, P., Rey, F., Schwarz, M., Walker, L.R., 2014. Ecological mitigation of hillslope instability: ten key issues facing researchers and practitioners. Plant Soil 377, 1–23. https://doi.org/10.1007/s11104-014-2044-6.

Stolpe, N.B., Hepp, C.K., Sandoval, M.E., Stuardo, R., Rodríguez, I., Almonacid, P., Ramírez, M., 2014. Caracterización taxonómica de los suelos de los valles de interés agropecuario de la Región de Aysén: (Patagonia Occidental-Chile), 300th ed. Imprenta América, Temuco, Chile.

Streck, C., 2020. Who Owns REDD+? Carbon Markets, Carbon Rights and Entitlements to REDD+ Finance. Forests 11, 959. https://doi.org/10.3390/f11090959.

Stubenrauch, J., 2022. Governance Analysis - Existing Regulations and Their Effectiveness, in: Stubenrauch, J., Ekardt, F., Hagemann, K., Garske, B. (Eds.), Forest Governance: Overcoming Trade-Offs between Land-Use Pressures, Climate and Biodiversity Protection (Volume 3), 1st ed. Springer International Publishing; Imprint Springer, Cham.

Sullivan, S., 2017. What's ontology got to do with it? On nature and knowledge in a political ecology of the 'green economy'. Journal of Political Ecology 24, 217–242. https://doi.org/10.2458/v24i1.20802.

Szymczak, S., Backendorf, F., Bott, F., Fricke, K., Junghänel, T., Walawender, E., 2022. Impacts of Heavy and Persistent Precipitation on Railroad Infrastructure in July 2021: A Case Study from the Ahr Valley, Rhineland-Palatinate, Germany. Atmosphere 13, 1118. https://doi.org/10.3390/atmos13071118.

Taki, H., Inoue, T., Tanaka, H., Makihara, H., Sueyoshi, M., Isono, M., Okabe, K., 2010. Responses of community structure, diversity, and abundance of understory plants and insect assemblages to thinning in plantations. Forest Ecology and Management 259, 607–613. https://doi.org/10.1016/j.foreco.2009.11.019.

Tanner, A., Árvai, J., 2018. Perceptions of Risk and Vulnerability Following Exposure to a Major Natural Disaster: The Calgary Flood of 2013. Risk analysis : an official publication of the Society for Risk Analysis 38, 548–561. https://doi.org/10.1111/risa.12851.

Taubenböck, H., Post, J., Roth, A., Zosseder, K., Strunz, G., Dech, S., 2008. A conceptual vulnerability and risk framework as outline to identify capabilities of remote sensing. Nat. Hazards Earth Syst. Sci. 8, 409–420. https://doi.org/10.5194/nhess-8-409-2008.

Tedim, F., Leone, V., Xanthopoulos, G., 2016. A wildfire risk management concept based on a social-ecological approach in the European Union: Fire Smart Territory. International Journal of Disaster Risk Reduction 18, 138–153. https://doi.org/10.1016/j.ijdrr.2016.06.005.

Temperton, V.M., Buchmann, N., Buisson, E., Durigan, G., Kazmierczak, Ł., Perring, M.P., Sá Dechoum, M. de, Veldman, J.W., Overbeck, G.E., 2019. Step back from the forest and step up to the Bonn Challenge: how a broad ecological perspective can promote successful landscape restoration. Restor Ecol 27, 705–719. https://doi.org/10.1111/rec.12989.

Terwilliger, V.J., 1990. Effects of vegetation on soil slippage by pore pressure modification. Earth Surface Processes and Landforms 15, 553–570. https://doi.org/10.1002/esp.3290150607.

Thiel, A., Moser, C., 2019. Foundational Aspects of Polycentric Governance: Overarching Rules, Social-Problem Characteristics, and Heterogeneity, in: Thiel, A., Garrick, D., Blomquist, W.A. (Eds.), Governing complexity: Analyzing and applying polycentricity. Cambridge University Press, Cambridge, United Kingdom, New York, pp. 65–90.

Thiel, A., Mukhtarov, F., Zikos, D., 2015. Crafting or designing? Science and politics for purposeful institutional change in Social–Ecological Systems. Environmental Science & Policy 53, 81–86. https://doi.org/10.1016/j.envsci.2015.07.018.

Thurner, M., Beer, C., Santoro, M., Carvalhais, N., Wutzler, T., Schepaschenko, D., Shvidenko, A., Kompter, E., Ahrens, B., Levick, S.R., Schmullius, C., 2014. Carbon stock and density of northern boreal and temperate forests. Global Ecology and Biogeography 23, 297–310. https://doi.org/10.1111/geb.12125.

Till-Bottraud, I., Fajardo, A., Rioux, D., 2012. Multi-stemmed trees of Nothofagus pumilio second-growth forest in Patagonia are formed by highly related individuals. Annals of botany 110, 905–913. https://doi.org/10.1093/aob/mcs146.

Tomasevic, J.A., Estades, C.F., 2008. Effects of the structure of pine plantations on their "softness" as barriers for ground-dwelling forest birds in south-central Chile. Forest Ecology and Management 255, 810–816. https://doi.org/10.1016/j.foreco.2007.09.073.

Toro-Manríquez, M., Soler, R., Lencinas, M.V., Promis, Á., 2019. Canopy composition and site are indicative of mineral soil conditions in Patagonian mixed Nothofagus forests. Annals of Forest Science 76, 1–14. https://doi.org/10.1007/s13595-019-0886-z.

Tröger, D., Braun, A.C. "Industry impacts more than nature: Overcoming Epistemic Challenges to Understanding the Risk Perception of Natural Hazards and Establishing Symmetrical Relations with an Actor-Network-Theory Approach". Int J Disaster Risk Sci.

Tröger, D., Braun, A.C., Eichel, J., Schmidtlein, S., Sandoval Estrada, M., Valdés Durán, A., 2022. Pinus plantations impact hillslope stability and decrease landscape resilience by changing biogeomorphic feedbacks in Chile. CATENA 216, 106364. https://doi.org/10.1016/j.catena.2022.106364.

Truedinger, A.J., Jamshed, A., Sauter, H., Birkmann, J., 2023. Adaptation after Extreme Flooding Events: Moving or Staying? The Case of the Ahr Valley in Germany. Sustainability 15, 1407. https://doi.org/10.3390/su15021407.

Turner, B.L., Kasperson, R.E., Matson, P.A., McCarthy, J.J., Corell, R.W., Christensen, L., Eckley, N., Kasperson, J.X., Luers, A., Martello, M.L., Polsky, C., Pulsipher, A., Schiller, A., 2003. A framework for vulnerability analysis in sustainability science. Proceedings of the National Academy of Sciences of the United States of America 100, 8074–8079. https://doi.org/10.1073/pnas.1231335100.

Uekötter, F., 2018. Techniker an der Macht. Der Ingenieur-Politiker im 20. Jahrhundert. Historische Zeitschrift 306, 396–423. https://doi.org/10.1515/hzhz-2018-0010.

Underwood, A.J., 1997. Experiments in ecology: their logical design and interpretation using analysis of variance. Cambridge University Press, Cambridge.

Valdés Durán, A., Velásquez, J., Neculqueo, G., Deckart, K., Lillo, D.C., Frez, L.N., Tröger, D., Quezada, A.C., Solé, M.B., Escobar, M.E., 2022. Extreme climatic events in northern Chile and their impact on the geochemical composition of the Huasco River. Journal of South American Earth Sciences 118, 103927. https://doi.org/10.1016/j.jsames.2022.103927.

Valdivieso, P., Andersson, K.P., 2017. Local Politics of Environmental Disaster Risk Management. The Journal of Environment & Development 26, 51–81. https://doi.org/10.1177/1070496516685369.

Valdivieso, P., Públicas, C., 2019. Institutional Drivers of Disaster Risk Reduction: A Comparative Study of Local Government Decisions and Outcomes in Chile.

Valenzuela, P., Arellano, E.C., Burger, J., Oliet, J.A., Perez, M.F., 2018. Soil conditions and sheltering techniques improve active restoration of degraded Nothofagus pumilio forest in Southern Patagonia. Forest Ecology and Management 424, 28–38. https://doi.org/10.1016/j.foreco.2018.04.042.

van Assche, K., Birchall, J., Gruezmacher, M., 2022. Arctic and northern community governance: The need for local planning and design as resilience strategy. Land Use Policy 117, 106062. https://doi.org/10.1016/j.landusepol.2022.106062.

van Daele, M., Versteeg, W., Pino, M., Urrutia, R., Batist, M. de, 2013. Widespread deformation of basin-plain sediments in Aysén fjord (Chile) due to impact by earthquake-triggered, onshore-generated mass movements. Marine Geology 337, 67–79. https://doi.org/10.1016/j.margeo.2013.01.006.

van der Sluis, T., Pedroli, B., Kristensen, S.B., Lavinia Cosor, G., Pavlis, E., 2016. Changing land use intensity in Europe – Recent processes in selected case studies. Land Use Policy 57, 777–785. https://doi.org/10.1016/j.landusepol.2014.12.005.

van Galen, L.G., Lord, J.M., Orlovich, D.A., Jowett, T., Larcombe, M.J., 2023. Barriers to seedling establishment in grasslands: Implications for Nothofagus forest restoration and migration. Journal of Applied Ecology 60, 291–304. https://doi.org/10.1111/1365-2664.14331.

van Holt, T., Crona, B., Johnson, J.C., Gelcich, S., 2017. The consequences of landscape change on fishing strategies. The Science of the total environment 579, 930–939. https://doi.org/10.1016/j.scitotenv.2016.10.052.

van Holt, T., Moreno, C.A., Binford, M.W., Portier, K.M., Mulsow, S., Frazer, T.K., 2012. Influence of landscape change on nearshore fisheries in southern Chile. Glob Change Biol 18, 2147–2160. https://doi.org/10.1111/j.1365-2486.2012.02674.x.

van Noordwijk, M., Duguma, L.A., Dewi, S., Leimona, B., Catacutan, D.C., Lusiana, B., Öborn, I., Hairiah, K., Minang, P.A., 2018. SDG synergy between agriculture and forestry in the food, energy, water and income nexus: reinventing agroforestry? Current Opinion in Environmental Sustainability 34, 33–42. https://doi.org/10.1016/j.cosust.2018.09.003.

van Riet, G., 2021. The Nature–Culture Distinction in Disaster Studies: The Recent Petition for Reform as an Opportunity for New Thinking? Int J Disaster Risk Sci 12, 240–249. https://doi.org/10.1007/s13753-021-00329-7.

van Well, L., van der Keur, P., Harjanne, A., Pagneux, E., Perrels, A., Henriksen, H.J., 2018. Resilience to natural hazards: An analysis of territorial governance in the Nordic countries. International Journal of Disaster Risk Reduction 31, 1283–1294. https://doi.org/10.1016/j.ijdrr.2018.01.005.

Vargas, G., Rebolledo, S., Sepúlveda, S.A., Lahsen, A., Thiele, R., Townley, B., Padilla, C., Rauld, R., Herrera, M.J., Lara, M., 2013. Submarine earthquake rupture, active faulting and volcanism along the major Liquiñe-Ofqui Fault Zone and implications for seismic hazard assessment in the Patagonian Andes. Andean Geology 40, 141–171. https://doi.org/10.5027/andgeoV40n1-a07.

Veblen, T.T. (Ed.), 1996. The ecology and biogeography of Nothofagus forests. Yale Univ. Press, New Haven, 403 pp.

Vergara, P.M., Simonetti, J.A., 2004. Avian responses to fragmentation of the Maulino Forest in central Chile. Oryx 38, 383–388. https://doi.org/10.1017/S0030605304000742.

Vesterdal, L., Clarke, N., Sigurdsson, B.D., Gundersen, P., 2013. Do tree species influence soil carbon stocks in temperate and boreal forests? Forest Ecology and Management 309, 4–18. https://doi.org/10.1016/j.foreco.2013.01.017.

Vicencio, K.M., Rodríguez, Alejandro, Vallina, Vicencio, J.G., 2023. El desafío de la descentralización administrativa en materia de ordenación territorial hacia los Gobiernos Regionales en Chile. An. geogr. Univ. Complut. 43, 359–383. https://doi.org/10.5209/aguc.90580.

Viles, H., 2020. Biogeomorphology: Past, present and future. Geomorphology 366, 106809. https://doi.org/10.1016/j.geomorph.2019.06.022.

Vorogushyn, S., Apel, H., Kemter, M., Thieken, A., 2022. Statistical and hydraulic analysis of flood hazard in the Ahr valley, Germany considering historical floods.

Wachinger, G., Renn, O., Begg, C., Kuhlicke, C., 2013. The risk perception paradox--implications for governance and communication of natural hazards. Risk analysis : an official publication of the Society for Risk Analysis 33, 1049–1065. https://doi.org/10.1111/j.1539-6924.2012.01942.x.

Walker, B., Salt, D., Reid, W., 2012. Resilience thinking: Sustaining ecosystems and people in a changing world. Island Press, Washington, 174 pp.

Wang, X., Ma, C., Wang, Y., Wang, Y., Li, T., Dai, Z., Li, M., 2020. Effect of root architecture on rainfall threshold for slope stability: variabilities in saturated hydraulic conductivity and strength of root-soil composite. Landslides 17, 1965–1977. https://doi.org/10.1007/s10346-020-01422-6.

Ward, P.J., Blauhut, V., Bloemendaal, N., Daniell, J.E., Ruiter, M.C. de, Duncan, M.J., Emberson, R., Jenkins, S.F., Kirschbaum, D., Kunz, M., Mohr, S., Muis, S., Riddell, G.A., Schäfer, A., Stanley, T., Veldkamp, T.I.E., Winsemius, H.C., 2020. Review article: Natural hazard risk assessments at the global scale. Nat. Hazards Earth Syst. Sci. 20, 1069–1096. https://doi.org/10.5194/nhess-20-1069-2020.

Watson, A., Phillips, C., Marden, M., 1999. Root strength, growth, and rates of decay: root reinforcement changes of two tree species and their contribution to slope stability. Plant and Soil 217, 39–47.

Watson, J.E.M., Shanahan, D.F., Di Marco, M., Allan, J., Laurance, W.F., Sanderson, E.W., Mackey, B., Venter, O., 2016. Catastrophic Declines in Wilderness Areas Undermine Global Environment Targets. Current biology: CB 26, 2929–2934. https://doi.org/10.1016/j.cub.2016.08.049.

Weber, T.K.D., Ingwersen, J., Högy, P., Poyda, A., Wizemann, H.-D., Demyan, M.S., Bohm, K., Eshonkulov, R., Gayler, S., Kremer, P., Laub, M., Nkwain, Y.F., Troost, C., Witte, I., Reichenau, T., Berger, T., Cadisch, G., Müller, T., Fangmeier, A., Wulfmeyer, V., Streck, T., 2022. Multi-site, multi-crop measurements in the soil–vegetation–atmosphere continuum: a comprehensive dataset from two climatically contrasting regions in southwestern Germany for the period 2009–2018. Earth Syst. Sci. Data 14, 1153–1181. https://doi.org/10.5194/essd-14-1153-2022.

Weichselgartner, J., Pigeon, P., 2015. The Role of Knowledge in Disaster Risk Reduction. Int J Disaster Risk Sci 6, 107–116. https://doi.org/10.1007/s13753-015-0052-7.

Wicki, A., Lehmann, P., Hauck, C., Seneviratne, S.I., Waldner, P., Stähli, M., 2020. Assessing the potential of soil moisture measurements for regional landslide early warning. Landslides 17, 1881–1896. https://doi.org/10.1007/s10346-020-01400-y.

Williams, M.C., Wardle, G.M., 2005. The invasion of two native Eucalypt forests by Pinus radiata in the Blue Mountains, New South Wales, Australia. Biological Conservation 125, 55–64. https://doi.org/10.1016/j.biocon.2005.03.011.

Wilson, M., Lovell, S., 2016. Agroforestry—The Next Step in Sustainable and Resilient Agriculture. Sustainability 8, 574. https://doi.org/10.3390/su8060574.

Wilson, T., Cole, J., Johnston, D., Cronin, S., Stewart, C., Dantas, A., 2012. Short- and long-term evacuation of people and livestock during a volcanic crisis: Lessons from the 1991 eruption of Volcan Hudson, Chile. Journal of Applied Volcanology 1, 2. https://doi.org/10.1186/2191-5040-1-2.

Wilson, T.M., Cole, J.W., Stewart, C., Cronin, S.J., Johnston, D.M., 2011. Ash storms: Impacts of wind-remobilised volcanic ash on rural communities and agriculture following the 1991 Hudson eruption, southern Patagonia, Chile. Bull Volcanol 73, 223–239. https://doi.org/10.1007/s00445-010-0396-1.

Windisch, M., Humpenoeder, F., Merfort, L., Bauer, N., Dietrich, J.P., Lotze-Campen, H., Seneviratne, S., Popp, A., 2023. Defending climate targets under threat of forest carbon impermanence. Copernicus Meetings EGU23–15161. Copernicus Meetings. https://meetingorganizer.copernicus.org/EGU23/EGU23-15161.html.

Wisner, B., Blaikie, P., Cannon, T., Davis, I., 2010. At risk: Natural hazards, people's vulnerability and disasters, 2nd ed. Routledge, London, 471 pp.

Zamorano-Elgueta, C., Moreno, P.C., 2021. Restoration for Provision of Ecosystem Services in Patagonia-Aysén, Chile, in: Peri, P.L., Pastur, G.M., Nahuelhual, L. (Eds.), Ecosystem Services in Patagonia: A Multi-Criteria Approach for an Integrated Assessment. Springer, Cham.

Zhang, D., Stanturf, J., 2008. Forest Plantations, in: Jørgensen, S.E. (Ed.), Encyclopedia of ecology, 1st ed. Elsevier, Amsterdam, pp. 1673–1680.

Zinngrebe, Y., Borasino, E., Chiputwa, B., Dobie, P., Garcia, E., Gassner, A., Kihumuro, P., Komarudin, H., Liswanti, N., Makui, P., Plieninger, T., Winter, E., Hauck, J., 2020. Agroforestry governance for operationalising the landscape approach: connecting conservation and farming actors. Sustain Sci 15, 1417–1434. https://doi.org/10.1007/s11625-020-00840-8.

Printed in the United States
by Baker & Taylor Publisher Services